Advances in Asian Human-Environmental Research

The *Advances in Asian Human-Environmental Research* series aims at fostering the discussion on the complex relationships between physical landscapes, natural resources, and their modification by human land use in various environments of Asia. It is widely acknowledged that human-environmental interactions become increasingly important in Area Studies and development research, taking into account regional differences as well as bio-physical, socio-economic and cultural particularities. The book series seeks to explore theoretic and conceptual reflection on dynamic human-environment systems applying advanced methodology and innovative research perspectives. The main themes of the series cover urban and rural landscapes in Asia. Examples include topics such as land and forest degradation, vulnerability and mitigation strategies, natural hazards and risk management concepts, environmental change, impact studies and consequences for local communities. The relevant themes of the series are mainly focused on geographical research perspectives of Area Studies, however there is scope for interdisciplinary contributions from various spheres within the natural sciences, social sciences and humanities. Key themes: Human-Environment Interaction – Asian regional studies – Asian geography – Impact studies – Landscape – Society – Land use and land cover change – Natural resources *Submit a proposal*: Proposals for the series will be considered by the Series Editor and International Editorial Board. An initial author/editor questionnaire and instructions for authors can be obtained from the Publisher, Dr. Robert K. Doe (robert.doe@springer.com).

Stanzin Passang

Snow Cover Distribution and Dynamics

The Trans-Himalayan Region of Ladakh, India

Springer

Stanzin Passang
Department of Geography
South Asia Institute, Heidelberg University
Heidelberg, Germany

ISSN 1879-7180 ISSN 1879-7199 (electronic)
Advances in Asian Human-Environmental Research
ISBN 978-3-031-57691-1 ISBN 978-3-031-57692-8 (eBook)
https://doi.org/10.1007/978-3-031-57692-8

This Springer imprint is published by the registered company Springer Nature Switzerland AG
The registered company address is: Gewerbestrasse 11, 6330 Cham, Switzerland

For the People of Ladakh

"Padum Castle"

Spring, oh, the three months of spring, are browner than agate.
The Padum castle of the lord of the world is the colour of agate.
The castle of Dharma King Nyima Namgyal is the colour of agate.

Summer, oh, the three months of summer, are greener than turquoise.
The Padum castle of the lord of the world is the colour of turquoise.
The castle of Dharma King Nyima Namgyal is the colour of turquoise.

Autumn, oh, the three months of autumn, are yellower than gold.
The Padum castle of the lord of the world is the colour of gold.
The castle of Dharma King Nyima Namgyal is the colour of gold.

Winter, oh, the three months of winter, are whiter than sacred seashells.
The Padum castle of the lord of the world is the colour of sacred seashells.
The castle of Dharma King Nyima Namgyal is the colour of sacred seashells.

(Traditional folk song in praise of the seasons at Padum, Zangskar)

Preface

This book is the result of my PhD research that was made possible by the funding provided by Henk Thoma and his family members from the Netherlands, and a grant provided by DAAD – Stipendien- und Betreuungsprogramm (STIBET) – Grant for the completion of a doctoral degree through Graduate Academy, Heidelberg University.

First and foremost, I would like to thank Prof. Dr. Marcus Nüsser for giving the opportunity and having faith in me to work on such an important research topic, which was missing in the region of Trans-Himalayan Ladakh. His constant support, motivation, and full trust throughout my doctoral research encouraged me to stay focused during the completion of this work.

I am highly thankful to Dr. Susanne Schmidt for her kind guidance and advice, especially in the field of remote sensing and GIS, which helped me navigate the course of the study on the right track, from the beginning till the end.

I am highly obliged to His Eminence Togdan Rinpoche for his blessing and lending me his four-wheel drive vehicle for the reconnaissance of initial field survey sites, and I owe immense gratitude to Venerable Lama Lobzang Angchuk for providing further transportation for snow sampling at high altitude sites, no matter the weather conditions.

I owe a depth of gratitude to Thupstan Motup, Rinchen Namgyal, Jigmet Tsewang, and Phuntsog Wangdus for accompanying me during the field data collection. Without their help, the task of field surveys would not have been easy, in an inhospitable high altitude and wilderness environment.

I would like to extend my appreciation to Blaise Humbert-Droz, Sonam Tsetan, and others who helped me directly or indirectly in many ways to get into the academic career.

The number of people who have helped me with smooth settling, initially in finding accommodation, and with administrative and other works in Heidelberg are many. I am grateful to all but can't mention all here. However, I wholeheartedly thank Dr. Johannes Anhorn, Dr. Ravi Baghel, Dr. Aseem Shrivastava, Micheal Ebner, Benjamin Kraus, Dr. Judith Müller, Dagmar Brombierstäudl, and Stefanie Raschke. And a profound thanks in particular to Dr. Claudia Breuer and the Ball family for giving me a warm home away from home experience during the final phase of my thesis "writing up."

To the current and past colleagues of South Asian Institute, Geography group, I would like to thank you all for creating a comfortable and enjoyable work environment. I thoroughly enjoyed the fruitful group meeting discussions and your constructive inputs into promoting my doctoral work. It has improved my overall personality over the last few years.

Most of all, I am highly indebted to my family, who has been selflessly supporting me and allowing me to stay far away from home to pursue a PhD.

This journey would have been incomplete without the company of great friends in Heidelberg, whom I owe a great deal for their permanent moral support, encouragement, patience, and sharing of views, to be able to keep on going forward in my life.

Heidelberg, Germany Stanzin Passang

Summary

Snow cover dynamics play an important role in the sustenance of glacier repositories in the Trans-Himalayan region of Ladakh, India. Understanding the changing nature of snow distribution helps assessing the possible impact of climate change, avalanches, forecasting, and various other applications.

The Union Territory of Ladakh is a high mountain region that to a large extent relies on irrigated agriculture for its food security. Due to the semi-arid climate conditions, the cryosphere of Ladakh serves as storage for water used for irrigation. Besides glaciers and permafrost, snow cover is one of the vital sources that ensures the availability of freshwater from seasonal snow melt.

This study investigated the distribution of the seasonal snow cover in the region of Ladakh, in the observation period from 2000 until 2019. The snow cover extent and duration were examined and mapped using Moderate Resolution Imaging Spectroradiometer (MODIS) daily snow cover products. The appropriate Normalised Difference Snow Index (NDSI) value representing snow in the region was established at 0.40, validated using Landsat 8 Operational Land Imager (LC8 OLI) cloud-free snow cover maps as a relative ground truth method.

Several algorithms were developed to be used in sequence that reduced the clouds from the MODIS satellite imagery to 3%, which helped to bridge the discontinuity in spatial and temporal scale for the daily long-term snow cover monitoring.

The monthly, seasonal, and annual MODIS snow cover area (SCA) and snow cover days (SCD) were examined up to a regional and subregional pixel-wise level, using different statistical approaches like linear regression, Mann-Kendall (MK) with a two-variance correction, Sen's slope trend, and Pearson's r and Spearman's ρ correlation coefficient tests. The regional snow line elevation (RSLE) on a regional level and the topographical impact (the elevation, slope, and aspect) on the snow cover distribution on a subregional level were analysed in more detail as well. A fitting forecast model for the prediction of snowfall in the region was also developed.

At a local scale, two custom-made weather stations and a snowboard were installed, and in situ snow surveys measuring snow depth and snow water equivalent (SWE) were conducted, in order to generate outdoor reference data.

Included as a separate case study, the installation of four terrestrial cameras in the Leh Valley, featuring different elevations, slopes, and aspects, facilitated a high resolution quantification and validation of the MODIS and Landsat 8 OLI snow cover distribution patterns and snow lines with the camera oblique photographs, orthorectified using the WSL Monoplotting Tool of the Swiss Federal Institute for Forest, Snow and Landscape Research. Repeat photography is relatively low-cost and without the need for direct field access a useful ground truthing method.

Overall, the SCA in Ladakh shows an annual mean of about 42%, ranging from about 15% in August to 71% in February, and displays a high variety between and within the subregions. Trends of the SCA display a non-significant minor increase in the summer as well as annually, but a non-significant decrease for spring and the winter. The month of July manifests a significant positive trend of the SCA of 0.3% year^{-1}. The regional snow line elevation shows a significant downward shift in June (−5.20 m year^{-1}) and July (−5.56 m year^{-1}), with non-significant positive trends only in the winter.

The average SCD of the entire study area is 151 days. The analysis of the SCD trend shows more pixels with a significant increase than decrease. Annually and in summer, 6% and 12%, respectively, of all pixels indicate a significant positive trend, and 2% a significant negative trend both annually and in winter. In the subregions Suru and Zangskar, 36% and 18%, respectively, of all pixels have an increasing trend during the summer.

The comparison amongst the coarse resolution MODIS, high Landsat, and very high resolution orthorectified photographs of the terrestrial camera stations shows a good agreement of the snow cover extent and snow lines.

The results of the dissertation are important for any hydrological and development studies against the backdrop of global warming and climate change.

Kurzzusammenfassung

Die Schneedecke spielt eine wichtige Rolle für die Massenbilanzen der Gletscher in der Trans-Himalaya-Region von Ladakh, Indien. Das Verständnis der sich verändernden Schneedeckendynamik hilft bei der Beurteilung der möglichen Auswirkungen des Klimawandels sowie bei der Beurteilung des Lawinenrisikos und Fragen der Wasserverfügbarkeit.

Das Unionsterritorium Ladakh ist eine Hochgebirgsregion, die in hohem Maße auf die Bewässerungslandwirtschaft angewiesen ist. Aufgrund der semiariden Klimabedingungen stellt die Kryosphäre eine wichtige Wasserressource dar, deren Schmelzwasser für die Bewässerung verwendet wird. Neben Gletschern und Permafrost stellt die saisonale Schneedecke eine der wichtigsten Wasserquellen dar.

In dieser Studie wurde die Verteilung der saisonalen Schneedecke in der Region Ladakh im Beobachtungszeitraum von 2000 bis 2019 untersucht. Die Ausdehnung und Dauer der saisonalen Schneedecke wurden auf Basis der täglichen Schneedeckenprodukte des Moderate Resolution Imaging Spectroradiometer (MODIS) untersucht. Unter Anwendung eines Grenzwertverfahrens aus den Normierten Differenz Schnee Index-Daten (NDSI) wurde die Schneedecke kartiert. Die Validierung des Grenzwertes erfolgte über ausgewählte wolkenfreie Landsat 8-Satellitenbilder (Landsat Operational Land Imager).

Zur Reduktion der Wolkenbedeckung wurden verschiedene Algorithmen nacheinander eingesetzt. Mit diesem Verfahren konnte die Wolkenbedeckung auf 3 % reduziert werden, womit zeitliche und räumliche Diskontinuitäten überbrückt wurden.

Die monatlichen, saisonalen und jährlichen MODIS-Schneedeckenflächen (SCA) und Schneedeckentage (SCD) wurden auf regionaler und subregionaler Pixelebene untersucht. Für Trendanalysen wurden der Mann-Kendall (MK)-Test, Sen's Slope Trend, Pearson's r und Spearman's ρ Korrelationskoeffizienten eingesetzt. Die regionale Schneelinienhöhe (RSLE) und der topographische Einfluss (Höhe, Neigung und Ausrichtung) auf die Schneedeckenverteilung wurden analysiert. Ein Vorhersagemodell der zukünftigen Entwicklung der Schneedecke wurde angewandt.

Zwei speziell angefertigte Wetterstationen und ein Snowboard wurden zur Messung der Schneehöhe und des Schneewasseräquivalent (SWE) installiert. Für eine weitere Validierung der aus den MODIS- und Landsat 8 OLI Daten generierten Schneedecken wurden vier terrestrische Kameras im oberen Einzugsgebiet von Leh installiert. Diese haben in unterschiedlichen Höhenlagen, Hänge und Expositionen mit einer hohen zeitlichen und räumlichen Auflösung aufgenommen. Die Bilder wurden mit dem WSL Monoplotting Tool der Eidgenössischen Forschungsanstalt für Wald-, Schnee- und Landschaftsforschung (WSL) ausgewertet. Terrestrische Wiederholungsaufnahmen sind relativ kostengünstig und ohne die Notwendigkeit eines direkten Feldzugangs eine nützliche Methode der Datenvalidierung.

Die Schneebedeckung in Ladakh weist einen Jahresmittelwert von etwa 42 % auf, der zwischen 15 % im August bis 71 % im Februar variiert. Insgesamt zeigen die Ergebnisse die große Variabilität der Schneedeckenverteilung zwischen und innerhalb der Teilregionen auf. Die Trendanalysen des SCA belegen einen nicht signifikanten leichten Anstieg im Sommer sowie im Jahresmittel, dagegen einen nicht signifikanten Rückgang im Frühjahr und Winter. Der Monat Juli zeigt einen signifikanten positiven Trend des SCA von 0,3 % yr^{-1}. Die regionale Schneelinienhöhe zeigt eine signifikante höhenwärtige Verschiebung nach unten im Juni ($-5{,}20$ m yr^{-1}) und Juli ($-5{,}56$ m yr^{-1}) und nicht signifikante positive Trends im Winter.

Die durchschnittliche SCD im gesamten Untersuchungsgebiet beträgt 151 Tage. Die Analyse des SCD-Trends zeigt mehr Pixel mit einer signifikanten Zunahme als Abnahme. Jährlich und im Sommer zeigen 6 % bzw. 12 % aller Pixel einen signifikant positiven Trend. 2 % weisen einen signifikant negativen Trend sowohl im Jahr als auch im Winter auf. In den Teilregionen Suru und Zangskar haben 36 % bzw. 18 % aller Pixel einen steigenden Trend während des Sommers.

Der Vergleich zwischen den MODIS-Daten mit geringer räumlicher Auflösung, den Landsat-Daten mit mittlerer räumlicher Auflösung und den sehr hoch aufgelösten orthorektifizierten terrestrischen Bildern zeigt eine gute Übereinstimmung der Schneedeckenausdehnung und der Schneelinien.

Die Ergebnisse der vorliegenden Arbeit sind vor dem Hintergrund des globalen Klimawandels für weitere hydrologische und sozioökonomische Entwicklungsstudien relevant.

Contents

About the Author

Stanzin Passang is a Ph.D. graduate researcher of the South Asia Institute (SAI) of the Universität Heidelberg, Germany, examining the snow cover in the Trans-Himalayan region of Ladakh, India. He has a broad interest in the cryosphere of the area, including the glaciers, permafrost and aufeis (icing), and is much concerned about global warming, climate change, and recent environmental disasters in the Ladakh area, e.g., glacial lake outburst floods (GLOFs), cloud bursts, and excessive rainfall events. He received an MA in Geography from the Mahatma Gandhi University in Meghalaya, India, and a bachelor's degree in History from the Delhi University, India. He is Treasurer of the Sustainable Development Forum of Ladakh and currently working on a weather research project in Ladakh that utilises automatic weather stations and the usage of drones for high resolution mapping on a subregional scale.

Abbreviations

2D	Two-dimensional
3D	Three-dimensional
ACF	Autocorrelation Function
ADF	Augmented Dickey-Fuller
AIC	Akaike's Information Criterion
AIC_c	Corrected Akaike's Information Criterion
ALOS	Advanced Land Observing Satellite
AOI	Area of Interest
ARIMA	Autoregressive Integrated Moving Average
ARMA	Autoregressive Moving Average
ASTER	Advanced Spaceborne Thermal Emission and Reflection Radiometer
AWiFS	Advanced Wide Field Sensor
AWS	Automatic Weather Station
CCA	Cloud Cover Area
DAAC	Distributed Active Archive Center
DEM	Digital Elevation Model
DHCP	Dynamic Host Configuration Protocol
EOS	Earth Observing System
FOV	Field of View
GIMP	GNU Image Manipulation Program
GIS	Geographic Information System
GLOF	Glacial Lake Outburst Flood
GOES	Geostationary Operational Environmental Satellite
HD	High Definition
HDF	Hierarchical Data Format
HDF-EOS	Hierarchical Data Format-Earth Observing System
HMA	High Mountain Asia
HRTI-3	High Resolution Terrain Information-3
IDE	Integrated Development Environment
IMD	India Meteorological Department

IoT	Internet of Things
IP	Internet Protocol
IRS-LISS	Indian Remote Sensing Satellite Linear Imaging Self Scanning Sensor
JPL	Jet Propulsion Laboratory
KPSS	Kwiatkowski-Phillips-Schmidt-Shin
LC8 OLI	Landsat 8 Operational Land Imager
LCA	Land Cover Area
LIDAR	Light Detection and Ranging
m a.s.l.	Metres Above Sea Level
MCA	Mixed Cover Area
MK	Mann-Kendall
MODIS	Moderate Resolution Imaging Spectroradiometer
MPT	WSL Monoplotting Tool
MRT	MODIS Reprojection Tool
MSS	Multispectral Scanner System
NASA	National Aeronautics and Space Administration
NDSI	Normalised Difference Snow Index
NDVI	Normalised Difference Vegetation Index
NIR	Near-infrared
NGA	National Geospatial-Intelligence Agency
NOAA	National Oceanic and Atmospheric Administration
NSIDC	National Snow and Ice Data Center
OLI	Operational Land Imagery
PACF	Partial Autocorrelation Function
PCC	Pearson Correlation Coefficient
PM	Particulate Matter
QA	Quality Assessment
RGI	Randolph Glacier Inventory
RLLE	Regional Land Line Elevation
RPi	Raspberry Pi
RSLE	Regional Snow Line Elevation
RTC	Real-Time Clock
SAR	Synthetic Aperture Radar
SARIMA	Seasonal Autoregressive Integrated Moving Average
SCA	Snow Cover Area
SCD	Snow Cover Day
SD	Standard Deviation
SDHC	Secure Digital High Capacity
SNOMAP	Snow Mapping Algorithm
SNOWL	Snow Line Mapping Approach
Spatio-Tempo Combi	Spatio-temporal Combination
SPi	Sleepy Pi
SPOT	Systeme Probatoire d'Observation de la Terre
SRTM	Shuttle Radar Topography Mission

SSE	Sen's Slope Estimator
SSH	Secure Shell
SWE	Snow Water Equivalent
SWIR	Short-Wave Infrared
T-A Combi	Terra-Aqua Combination
TanDEM-X	TerraSAR Add-on Satellite for Digital Elevation Measurement
Tempo Combi	Temporal Combination
TIRS	Thermal Infrared Sensor
TM	Thematic Mapper
UNESCO	United Nations Educational, Scientific and Cultural Organization
USGS	United States Geological Survey
UV	Ultraviolet
VHR	Very-High Resolution
WMO	World Meteorological Organization
WSL	Swiss Federal Institute for Forest, Snow and Landscape Research

Chapter 1
Introduction

Abstract This chapter introduces the topic of the seasonal distribution of snow cover in the Trans-Himalayan region of Ladakh, India, during the years 2000–2019. The research delves into the consequences of shifting snow patterns and duration on the landscape, encompassing assessments of climate change and various applications involving satellite remote sensing, terrestrial imagery, on-site measurements, and weather data. The investigation also includes the testing of predictive models to anticipate how snow cover might change in the future, along with the optimal placement of camera stations. Given Ladakh's reliance on irrigated farming to secure its food supply in a region with limited rainfall, this study highlights the significance of the cryosphere – comprising glaciers, permafrost, and snow – as a critical water reservoir. Beyond categorising the terrain and detailing administrative aspects, this chapter provides insights into the local climate, socio-economic conditions, flora, and fauna. Through this study, we gain a deeper understanding of the intricate interplay between changing snow cover and Ladakh's distinct socio-environmental context, emphasising the indispensable role of the cryosphere in managing water resources.

Keywords Snow · Remote sensing · Terrestrial imagery · Snow courses · Leh · Ladakh · Trans-Himalaya · India

1.1 State of the Art

Snow cover is an important component of the cryosphere, derived from the Greek word Krios meaning "cold." The Earth's snowy and icy regions are collectively known as the cryosphere, and since snow is also a highly reflective material (with the reflectivity or albedo of fresh snow being 0.8–0.9), it plays a crucial role in the global climate system by modifying energy and moisture fluxes between the surface and the atmosphere and by acting as a water store in hydrological systems (Thackeray et al. 2019).

Snow is defined as falling or deposited ice particles formed mainly by sublimation, according to the World Meteorological Organization (WMO) and the

S. Passang, *Snow Cover Distribution and Dynamics*, Advances in Asian Human-Environmental Research, https://doi.org/10.1007/978-3-031-57692-8_1

United Nations Educational, Scientific, and Cultural Organization (UNESCO). The depth of fresh snow which falls during a given "recent" period is termed snowfall (Goodison and McKay 1978).

Measurement of snowfall is summed to determine the total for a single storm, a day, a month, or a year. Thus snow cover comprises the net accumulation of snow on the ground resulting from precipitation deposited as snowfall, ice pellets, hoar frost, and glaze ice, and water from rainfall, much of which has subsequently frozen at the time of an observation. The ground may be either completely or partly covered (Gray and Male 1981).

Snow cover is a key component of the global hydrological cycle (Hinkler et al. 2002; Barnett et al. 2005). The Himalayan, Karakoram, and Tibetan Plateau regions are known as the "Water Tower of Asia" or the third pole (Mukherji et al. 2015), as these mountain ranges occupy one of the highest concentrations of snow cover and ice outside the polar regions, and these high-altitude cryospheric components are unique as they are located in the subtropics rather than solely dominated by geographic latitude (Jain et al. 2009; Schmidt et al. 2009; Misra et al. 2020).

The Himalayan mountain ranges act as a climatic barrier between South and Central Asia, and their cryospheric components are essential for sustaining the freshwater storage for the many rivers originating in them (Immerzeel et al. 2013; Singh et al. 2014; Wani et al. 2021). Over the high-altitude region, the accelerated change in snow cover area (SCA) could be affecting glaciers along with the hydrological role pattern of the basin (Dharpure et al. 2020).

The monitoring and mapping of SCA and its characteristics plays an essential role in managing water resources for irrigation and drinking water, as well as understanding the impact of climate change over the region. The shift in SCA variability can affect river run-off and long-term seasonal snowmelt water supply. Therefore, quantification of these variations is of utmost significance for the sustenance of high-altitude repositories of frozen water components over the region (Wani et al. 2021).

Notwithstanding the importance of the seasonal snow cover, "High-altitude snowfall seems to be neglected and is obviously still rather unknown" (Winiger et al. 2005, p. 2329). In the Himalayan region, existing literature on in situ snow observation is sparsely sourced due to the complex topography, and to overcome this limitation, the monitoring of in situ snow field measurement is predominantly based on multi-temporal remote sensing data. In this regard, several studies used Thematic Mapper (TM), Landsat Multispectral Scanner System (MSS), Geostationary Operational Environmental Satellite (GOES), National Oceanic and Atmospheric Administration (NOAA), Indian Remote Sensing Satellite Linear Imaging Self Scanning Sensor (IRS-LISS) III & IV, Advanced Wide Field Sensor (AWiFS), and Systeme Probatoire d'Observation de la Terre (SPOT) (Hall 1995; Kulkarni et al. 2006).

Nonetheless, numerous studies revealed hindrances of these sensors regarding small swath, band resolution, and inaccuracy in the spatio-temporal assessment of SCA (Dankers and De Jong 2004; Kulkarni et al. 2006). In contrast to this, different snow products of Moderate Resolution Imaging Spectroradiometer (MODIS) have been used for daily, 8-day, and monthly global-gridded products to quantify the SCA and snow cover days (SCD), which show a higher degree of accuracy in different parts of the world (Hall and Riggs 2007; Jain et al. 2008; Gafurov and Bárdossy 2009b; Dong and Menzel 2016; Parajka et al. 2017). However, the application of MODIS snow cover products is affected by a high cloud obscuration, which causes a significant disruption in the uninterrupted long-term snow cover monitoring. Hence, there is a prerequisite to mitigate the cloud cover applying cloud removal approaches prior to the application of any snow characterisation. Numerous authors developed algorithms to de-cloud MODIS snow products based on spatial, temporal, and terrain information (Parajka and Blöschl 2008; Gafurov and Bárdossy 2009a; Tran et al. 2019).

Only a handful of studies have evaluated seasonal snow cover distribution, duration, and trends on a Himalayan scale (Immerzeel et al. 2009; Dharpure et al. 2021; Muhammad and Thapa 2021) and at a regional scale across the High Mountain Asia (HMA) on snow cover extent, snow depth, SCD, and the forecasting of snow (Jain et al. 2009; Gusain et al. 2014; Dai et al. 2017; Ahmad et al. 2018; Bilal et al. 2019; Dharpure et al. 2020; Misra et al. 2020; Saydi and Ding 2020; Kumar et al. 2022). However, the assessment of snow cover distribution and trends for the Himalayan region as a whole is intricate, as there is tremendous interregional contrast and high variation within a subregional scale (Dharpure et al. 2020). Therefore, precise studies are a requisite for the enhanced understanding of snow cover distribution and relevant changes. A limited number of studies deal with the role of topographical parameters such as elevation, slope, and aspect on the distribution of seasonal snow cover in the Himalayas (Jain et al. 2009; Dharpure et al. 2020; Sood et al. 2020).

The present-day in situ snow observation and snow stations are not available for the Ladakh region, and there is also limited knowledge of the normal state and historical trends. Many studies are focused on the fluctuation in glacier lengths, glacier-covered areas, and mass balance, for example, in Siachan (Bhutiyani 1999), Kang Yatse and Ladakh (Schmidt and Nüsser 2012, 2017), Upper Shayok Valley (Bhambri et al. 2013), Kang Yatse and Stok (Soheb et al. 2020, 2022), and the distribution and relevance of aufeis (icing) in the Upper Indus and Tshomoriri basin (Brombierstäudl et al. 2021, 2022). Snow research is still limited and relatively unexplored in the region of Ladakh, with over the past decade studies being confined to small-scale and short-term projects (Bhutiyani 1999; Ajai et al. 2010), and few researchers monitoring the SCA using MODIS over the entire Karakoram and Himalayan regions via remote sensing approaches. However, this was performed without any adequate ground truthing for the Ladakh region (Saydi and Ding 2020; Dharpure et al. 2021; Muhammad and Thapa 2021). Moreover, these studies directly applied the global reference Normalised Difference Snow Index (NDSI) threshold value over the whole Himalayan belt, which is in abundance of dense forest cover, where Zhang et al. (2019) suggested to set the threshold value as low as 0.01. SCA mapping validated with in situ measurements adopted an NDSI value of greater than or equal to 0.2 in Liaoning province, north-eastern China (Zhang et al. 2010), and a combination of Normalised Difference Vegetation Index (NDVI) and NDSI was used for the snow detection (Thapa et al. 2020).

Due to the scarcity of ground-based measurements and related historical snow data in the Ladakh region, it is extremely important to develop a detailed overview of the snow cover distribution and impacts of topographical factors on this snow cover, combined with field validation. Furthermore, to prepare a basis for assessing future trends, methods covering a detailed analysis of each basin scale, using different higher resolution data for the purpose of validating coarser remote sensing data such as the integration of time-lapse or repeat photography, can provide crucial information in data-scarce highly rugged mountain regions (Nüsser 2001; Schmidt 2009; Steiner 2011; Stockdale 2017; Nüsser and Schmidt 2021; Passang et al. 2022).

Snow cover dynamics play an important role in the sustenance of glacier repositories in the Himalayan regions. The glaciated area is very small, and cirque glaciers smaller than 1 km^2 are dominant in the Ladakh and Zangskar mountain ranges (Schmidt and Nüsser 2012; Soheb et al. 2022). Concern has been raised that global warming reduces the glaciers and their capacity to store water, as well as the amount of seasonal snow available for melting. Whether the melt water comes from snow or glacier ice, stable snowfall is required to maintain the flow in the long run (Bates et al. 2008).

To address the issue of the reduction and water storage capacity of glaciers, "artificial glaciers" represent an innovative intervention, strategically positioned at lower altitudes compared to the naturally occurring glaciers above 5200 m a.s.l., as noted by Schmidt and Nüsser (2017). These artificial glaciers can play a crucial role in mitigating water scarcity issues by serving as ice reservoirs that melt earlier in the agricultural season, thereby bridging the critical gap in water availability (Norphel and Tashi 2015; Nüsser and Baghel 2016; Nüsser et al. 2019a,b).

Besides glaciers and permafrost, snow cover is the primary vital water source that ensures the availability of water for the agricultural activities, groundwater table recharge, and other means of applications (Wani et al. 2021).

Due to the cold-arid climate conditions, the cryosphere of Ladakh serves as a storage for the water supply for households, wildlife, irrigation, hydro power generation, and some disservices such as avalanches, floods, and road blockages on high passes. Ladakh in the winter is isolated from the Indian plains, and as it is highly significant for food, trade, transportation, and the security of the nation, an assessment and monitoring of snow is essential (Chevuturi et al. 2018; Dame et al. 2019).

The livelihood of the local people directly depend on the snowmelt run-off, which is used to irrigate the fields. Even small shifts in the snow cover distribution will affect the local people (Schmidt and Nüsser 2017). However, major changes in the socio-economic scenario in Ladakh have taken place, as a result of increased income opportunities from non-agricultural activities provided by tourism, military, government, and other sectors (Nüsser et al. 2012; Dame 2018).

All these changes have intensified local water demands, and it has become an even more scarce commodity, especially around Leh town (Müller et al. 2020; Müller 2022). Water and water sources have also become increasingly more polluted. Effects on the snow cover and glaciers in the upper catchments are also observed, as, for example, vehicle air pollution (especially from diesel engines) settles on the snow and ice cover, influencing the snow albedo propelling faster snowmelt (Romshoo et al. 2022).

Immediate effects will be noticeable for the most vulnerable, the local farmers, most seriously farmers of the foothills, and lower downstream rivers. They rely on predictable water supplies to grow the crops that feed the nation in the mountains' shadows. Over one billion people rely on water resources originating from snowmelt run-off (Goodison et al. 1999).

For these reasons accurate snow monitoring and knowledge of its long term behaviour have become of vital importance. An early warning system of natural calamities due to snow should be a contributory factor in the policymakers' decisions, to act on firmly to reduce any potential dangers caused by snow. An example in Ladakh is the eastern and north-eastern areas of Changthang ("Northern Plains"), where livestock, especially Changra goats of the nomadic Changpa tribe, face death during winters with heavy snowfall events. In February 2008 some 150,000 Changra goats were at risk of death when the winter pastures got covered in about 60 cm of snow, the highest snowfall in decades (Bhasin 2012). And in March 2013 a period of heavy snow was followed by a shortage of fodder, killing almost 30,000 goats (Yatoo et al. 2014). Detection of critical snow conditions can be used in early warning systems for snow-related natural hazards (Scherer et al. 2005).

The negative change in winter snowfall is on the one hand affecting the glaciers and stream flow in Ladakh, witnessing much agricultural land being left barren due to a shortage of water. On the other hand, frequent rainfall causing flash floods and cloudbursts (Thayyen et al. 2013; Bhan et al. 2015; Ziegler et al. 2016), and the fast melting of snow cover and glaciers causing glacial lake outburst floods (GLOFs), has triggered destruction of agricultural fields, housing, infrastructure, and livestock (Schmidt et al. 2020; Majeed et al. 2021).

Therefore, compared to previous studies and against the backdrop of knowledge gaps, this work monitors the daily cloud-free SCA using an appropriate NDSI threshold value adjusted to the region, reducing MODIS cloud obstructions by combined spatial, backward–forward temporal, and regional snow line mapping approach filters. A detailed pixel-wise study was conducted to analyse the spatio-temporal patterns and trends of SCD and SCA for 19 snow hydrological years (2000–2019), using the non-parametric Mann–Kendall (MK) test (Mann 1945; Kendall 1948) and Sen's slope estimator (SSE) test (Sen 1968). Furthermore, time series and forecasting of snow to prepare a model basis for assessing future trends. Also, in situ field measurements of the snow depth and snow water equivalent (SWE) and a correlation analysis of the SCA, Regional Snow Line Elevation (RSLE), and climatic variables were performed.

Due to Ladakh's status as a sensitive border area, consistent meteorological data from weather stations is limited in the public domain. Therefore a custom-made automatic weather station (AWS) and a snowboard were installed, to generate reference data at 3660 metres above sea level (m a.s.l.). The station is equipped with two cameras and has been measuring air temperature, relative humidity, barometric pressure, soil temperature, wind direction, wind speed, and the ultraviolet (UV) index since spring 2016. Snow depth and SWE have been acquired from the snowboard on a daily basis since November 2015.

1.2 Objectives and Research Questions

- Forming the core of this research, to provide a detailed analysis of the seasonal snow cover distribution and duration in Ladakh, using daily MODIS snow cover product data for the observation period 2000–2019.
- To investigate the spatio-temporal patterns and trends of the snow-covered area and days in the region, delineated into seven subregions based on mountain ranges and watershed areas.
- To detect and quantify trends in the impact of topography (elevation, slope and aspect) on the SCA in these subregions.
- Furthermore, to develop a forecast model in order to make a prediction of the future snowfall trends at a regional scale.
- At a local scale, to perform in situ field observations including field surveys, the installation of four camera stations, and the building and installation of custom-made weather stations for validation purposes and to gather local weather data.
- Subsequently, to run an accuracy assessment and validation on the generated MODIS SCA, using satellite Landsat 8 Operational Land Imager (LC8 OLI) false colour snow composite data, referenced as ground truth and also, at a very high-resolution, a comparison of time-lapsed orthorectified photographs with high-resolution Landsat 8 OLI and coarse MODIS snow data, for the purpose of validation and delineation of the snow line in small watershed areas.

1.3 Study Area: The Trans-Himalayan Region of Ladakh

1.3.1 Extent

The study area encompasses an area of 65578 km^2 and extends from longitude 75°32'E to 79°46'E and latitude 32°34'N to 35°67'N. The elevation ranges from 2540 m a.s.l. in the lower Indus Valley near Kargil in the west to 7742 m a.s.l. on the Saltoro Ridge in the north. The mean elevation of Ladakh equals 4875 m a.s.l., which increases from the west to the east. It is bounded to Tibet from the north to the south-east, Baltistan to the north-west, the Kashmir Himalayas to the west, and the Lahaul and Spiti districts of Himachal Pradesh to the south. The Union Territory (since 31 October 2019) of Ladakh is nestled at the western edge of the Tibetan Plateau on the northern side of the Greater Himalayan range in India (see Fig. 1.1).

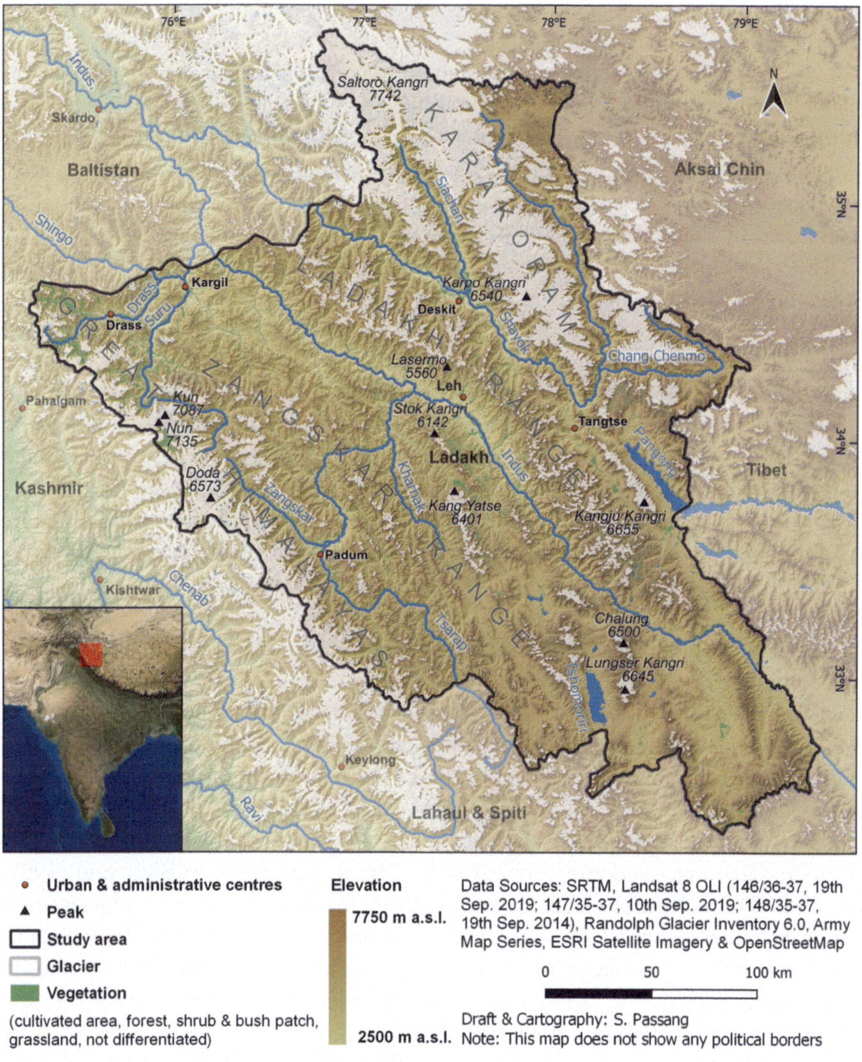

Fig. 1.1 Study area: The Trans-Himalayan region of Ladakh

Ladakh (Tibetan: ལ་དྭགས་ , Wylie:[1] *la dwags*) is a high mountain region in the Himalaya that to a large extent relies on irrigated agriculture for its food security. It is frequently translated as "land of passes," derived from the word "la" – pass; however "dakh" does not have a meaning on its own (Norman 2019). The Sanskrit

[1] Wylie is a method for the transliteration of Tibetan script into Roman script, created by the American scholar Turrell V. Wylie (Wylie 1959).

term Himalaya owes its name to the ancient pilgrims of India, meaning "abode of snow."

Ladakh lies athwart four major mountain ranges, which run in a general south-east to north-west direction and delimit Ladakh's main subregions. The Karakoram in the north/north-west and the Greater Himalayas in the south/south-west are the two major mountain ranges, which act as natural borders demarcated on the basis of watershed basins, whereas the Zangskar and Ladakh ranges run parallel through it, intersected by the flow of the Indus river within its boundaries.

The Indus, locally known as Singge Khabaps, originates on the Tibetan Plateau in the vicinity of Lake Manasarovar near Mount Kailash, flows north-west, and enters the Ladakh region at a place called Demchok in the Nyoma region of Changthang. Initially it flows between the Karakoram and the Ladakh range; however at a place called Dungti in the south-eastern region of the study area, the river takes a sharp south-western turn and cuts through the Ladakh range. It then again continues to follow a north-western course and flows towards the Leh region along the Ladakh range to the north and the Stok range to the south. After reaching Leh, the tributary Zangskar river joins the Indus near Nyemo village at the confluence called Sumdo. It continues its north-western course and reaches the town of Batalik, which is in Kargil district. Thereafter it enters into the Baltistan region, flowing through the town of Skardo.

The Indus follows partly the Indian continental plate that had drifted northwards and collided with the Eurasian continental plate. Geological evidence of this can be seen from the formations of igneous rocks, predominantly present to the north of the Indus, and sedimentary rock formations to the south of the Indus in the central Ladakh region. The upper Indus Valley marks the suture zone where the Indian plate is submerged under the Eurasian plate (Van Andel 1985; Philip and Mazari 2000).

The Karakoram range contributes about 3% of the total area of ice outside the ice-sheets in Greenland and Antarctica (Cogley 2012). According to Bhutiyani (1999, p. 113), "The Karakoram Himalaya is highly glaciated (about 37% glacier area) and is characterised by steep slopes and an altitude range of 2800 to 7600 m a.s.l." The Siachan glacier is one of the longest (approximately more than 70 km) glaciers in the world outside the Arctic and Antarctic polar regions and is situated in the Karakoram range, the north of the Nubra Valley.

The Shayok (also written Shyok) river runs from the Durbuk block region into the Nubra Valley and on to Baltistan between the Ladakh and the Karakoram mountain ranges. One of the prominent tributaries of the Shayok river is the Siachan river, originating from the Siachan glacier. In maps and guidebooks it is often addressed as "Nubra river"; however this name is categorically rejected by the local population (Norman 2019).

The Shayok and Siachan river valleys between the Ladakh range and the Karakoram bound Ladakh to the north, and the Suru and Zangskar river valleys between the Greater Himalayas and Zangskar range mark Ladakh's southern boundaries. The Indus Valley between the Zangskar and Ladakh ranges comprises a central territory, more or less cutting the area into two equal halves.

The eastern and north-eastern areas bordering Tibet predominantly comprise high-altitude cold windswept plains, rolling hills and wider valleys. This region, the Changthang, is considered to be an extension of the Tibetan Changthang into India (Humbert-Droz and Dawa 2004; Shukla and Srivastava 2020). In the south-eastern part of Ladakh, the parallelism of range and valley dissolves in a complex of high-altitude ridges and plateaus, many of them forming enclosed basins dotted with lakes and wetlands. Once covered by an extensive lake system, today only a few vestiges remain including Pangong Tsho, Tshomoriri, Tshokar, Kyun Tsho, Kyagar Tsho, and the extensive marshes near Hanle, Chumur, and Chushul (Pfister 2004).

Administratively the Union Territory of Ladakh is subdivided into two districts, named after the two main district capitals in the region, Leh and Kargil. Kargil district comprises the Zangskar (block headquarter at Padum) and Suru Valley regions, while the larger Leh district comprises the Indus and Nubra Valley regions (block headquarter at Deskit) and Changthang (block headquarter at Tangtse). The delineation of the area in this study does not follow any political borders and represents mountain ranges and watersheds.

1.3.2 Climate

The study area has extreme climatic conditions: Temperatures range from over 30 °C in summer to less than −40 °C in winter, and the region receives very low precipitation, which decreases eastwards from about 250 mm in Zangskar to less than 100 mm in the Indus Valley and Changthang regions (Dahri et al. 2016). The Greater Himalayan range acts as a barrier for the eastward moving moisture-laden wind system, and the amount of precipitation decreases considerably towards east of this range. Leh is situated north-east of the Zangskar range and Changthang in the eastern part of Ladakh receives a minimal amount of annual precipitation, as most of the weather system moves from west to east across the region (Bhan et al. 2015). The valleys, particularly those of Drass, Suru, and Zangskar, being just across the Greater Himalayas on the south-western part of the region, receive ample amount of winter precipitation in contrast to the other parts situated further east of the Zangskar and Ladakh ranges. These valleys are usually covered by heavy snow in the winter months, which allows moderate greening of the lower valley floors during the summer (Shafiq et al. 2016).

Snow cover has a significant influence on the weather and climatic patterns of Ladakh. The region falls under a rain shadow zone during the south-west monsoon in summer and under the influence of an extratropical low-pressure system known as the Western Disturbances in winter (Bhutiyani et al. 2010; Bhan et al. 2015; Chevuturi et al. 2018). As far as clouds, rain, and snowfall are concerned, the monsoon does not penetrate the Greater Himalayas much on the one hand, but on the other hand in the Karakoram higher elevations receive snowfall all year round, while the lower elevations receive 95% of their precipitation as snow and the remaining

5% as rain. The snow cover is relatively less and shallow, comparatively to the windward side of these mountain ranges (Bhutiyani 1999; Khan and Adams 2019).

Ladakh is predominantly characterised by fragile soil ecology at significant risk of degradation. Due to enhanced occurrences of extreme weather events and natural disasters, the number of conducted studies related to the climate in the region has increased rapidly, and it continues to receive a lot of media attention as well (Nüsser and Baghel 2014).

For example, Bhutiyani et al. (2007) revealed a significant rise in air temperature in the north-west Himalayan region, about 1.6 °C in the last century, with winters warming at a faster rate. Bhan et al. (2015) analysed monthly precipitations computed from the daily precipitation records of the meteorological observatory located at Leh airport for the period from 1901 to 2003 (excluding missing data of 21 years (1945–1947, 1970–1978, 1988, and 1990–1997)), obtained from the National Data Centre of the India Meteorological Department (IMD) of Pune and Srinagar, respectively. The results show that the total annual precipitation is about 105 mm, with 12.3 mm in July and 15.1 mm in August being the rainiest periods of mean monthly precipitation, followed by 9.1 mm in January and 8.8 mm in September (where 1 cm of snowfall is considered equal to 1 mm of liquid precipitation), similar to the findings of Dimri and Chevuturi (2014); Gusain et al. (2014). The highest ever accumulated rainfall in 24 hours is 51.3 mm, recorded on 22 August 1933 over Leh (IMD 1999). The heavy rainfall events of 5 August 2010 and 25 July 2011 have been found to be associated with cloudbursts during active monsoon conditions, due to westward moving cyclonic circulations in the middle troposphere over the region of Tibet-Ladakh (Bhan et al. 2015). Chevuturi et al. (2018) conducted a study to understand the climate of the Ladakh region and to characterise its changing climate used available weather data from several in situ climate stations. The research found that the Leh region has a cold and arid climate, with extreme temperature ranges from 34.8 °C in summer to −27.9 °C in winter and an annual mean temperature of 7.3 °C. Precipitation occurs mostly during the monsoon period, with cloudbursts also taking place during this time. Winter and pre-monsoon seasons experience significant precipitation due to the Western Disturbances. The average daily precipitation ranges from 0.5 to 1.5 mm/day. The study also concluded that the Ladakh region has experienced increasing temperature and decreasing precipitation trends during winter and summer periods in recent years. According to Shafiq et al. (2016), the winter precipitation shows a rising trend of nearly 0.04 mm per year in the form of snowfall, and the summer precipitation shows a decreasing trend with a reduction of 0.127 mm per year, usually in the form of rainfall, observed over the time period 1901–2000 in Leh.

From a socio-climatic perspective, weather-related information can be traced back to old traditional Ladakhi folk songs, for example, the song included in the front matter of this book, that is in praise of the four seasons at Padum, Zangskar during the reign of its king Nyima Namgyal (1694–1729). Local residents also relate to traditional monastic festivals in order to measure the relative temperature, which is conveyed in terms of the size of animals, e.g., a sheep during the Gustor festival in Spituk and a yak during the Dosmoche celebrations in Leh.

1.3.3 Socio-Economic Background

In western and central Ladakh, farming subsists on irrigation and is traditionally combined with livestock rearing, which includes short seasonal migrations to higher pastures (transhumance). The staple crops consist of barley, wheat, potatoes, peas, carrots, turnips, radish, green leafy vegetables, mustard, etc. Livestock provide food, fibre, skin for clothing, mode of ploughing the agricultural fields, and mode of transportation. Consuming fodders such as alfalfa, sparsely distributed vegetation, and crop residues, livestock produces manure which is used as organic fertilisers as well as fuel for cooking and heating (Dame and Nüsser 2008; Nüsser et al. 2012).

In contrast the nomadic herders (Changpa) inhabiting the high plateaus of eastern Ladakh undertake substantial migrations at different times of the year, making use of the available pasture resources. Goat, sheep, and yak husbandry is the mainstay of the local economy. The high-quality pashmina wool of the Changra goat, indigenous to the region, and wool and meat from the local breed of sheep and yak are the main cash products (Pfister 2004).

According to local elder views passed on from generation to generation, gradually improved connectivity over the last few decades, via land and air routes, made Ladakh slowly become a tourist hub, which took place in three distinguishable phases: the early phase, with the start of tourism in Ladakh in 1974. About 500 tourists visited Ladakh the first year. It was very difficult to visit in those days, and a Tibetan proverb best fits here: "If a valley is reached by a high and difficult pass, only the best of friends and worst of enemies are its visitors." Similarly foreign visitors who initially arrived were a positive novelty for Ladakh, as they brought first-hand exposure of the kind of industrialisation and modernisation as experienced in the west, and it spread awareness amongst local communities regarding the drawbacks of western-style development in the long run (Hodge 2013).

The second phase was a period of cultural tourism. A shift from local house to guest house and hotel was evident and with that an increase in the consumption of water during the peak tourist season, due to bathrooms fitted with running water and the replacement of local dry toilets with western-style flush toilets (Müller et al. 2020; Müller 2022).

The third phase started with the phenomenon of mass tourism, especially after a Bollywood movie, shot partly in Ladakh and released in 2009, became a big hit in India. This, in combination with a growing domestic economy with more prosperity and time for leisure, has attracted especially crowds of domestic origin to Ladakh. As per records registered at the state Tourism Department, 2018 saw a record of around 330.000 tourists visiting Ladakh, which outnumbered the total local population by around 50.000. Of these visitors about 50.000 were of foreign origin. Mass tourism has been seriously impacting the fragile environment of Ladakh, with construction booming in the form of a multiplication of guest houses and hotels, with their own boreholes for water consumption, as well as a drastic increase in the number of vehicles.

The sensitive disputed border situation with Pakistan and China has seen a constant increase in the influence of the military, with a growing defence infrastructure involving roads and installations (Baghel and Nüsser 2015).

A rural urban migration, predominantly to the city areas of Leh and Kargil, has caused the sprouting of new settlements, especially around Leh (Dame 2018). Also the shift from joint families to nuclear families has led to many new constructions, frequently on arable land.

Prosperity in the Leh region has seen many people buying up plots of lands, drilling boreholes for their water supply, and an enormous increase in vehicles for private use and status. Moreover, due to the boreholes, there is now less demand for surface water, causing a lack of maintenance of irrigation channels and ponds. The ground water table in the region has been negatively affected with a high contamination of heavy metals (Chorol and Gupta 2023), and lots of springs are drying up as well (Müller et al. 2020).

1.3.4 Flora and Fauna

Ladakh is situated at the confluence of two of the major biogeographical realms, the Palaearctic to the north and Indomalayan (oriental) to the south. It is endowed with a diverse and highly evolved flora and fauna (Humbert-Droz and Dawa 2004; Hartmann 2009; Doležal et al. 2018).

Flora

The Ladakh region belongs to the Trans-Himalayan vegetation zone which is dominated by desert steppes. Forests are virtually absent except for rare patches in locations where environmental conditions allow it (Dame and Nüsser 2008; Hartmann 2009; Kamp et al. 2011). The slope aspect affects the angle and duration of solar radiation, and consequently the amount of moisture in the soil on a north-facing slope is longer covered with snow than on a south-facing slope. Slope influences the radiation budget and water supply, both indicators of the vegetation. Since north-eastern and northern mountain slopes in Ladakh are longer covered with snow than south-western and southern mountain slopes, the vegetation cover on these slopes is denser than on the drier southern and western slopes. The region falls under a rain shadow area with little precipitation, and the intensity of the sun is really strong due to the thin air at the higher elevation of the terrain. Moreover, in the winter season the angle of the sun is more tilted, so snow on the northern mountain slopes remains mostly in the shadow. In this way there is also hardly sublimation of snow on these slopes during the winter time, and even in summer melting and sublimation of the northern exposed snow cover is less compared to the southern exposed snow cover (Downs et al. 1999; Khali et al. 2022).

Usually moisture-laden winds shade most of the moisture in the form of precipitation in the foothills of the Himalayas, which is on the windward side. That is why vegetation generally tends to be more prominent on southern mountain slopes. However in the case of Ladakh, the movement of wind circulation from west to east across the region, due to the orientation of the major mountain ranges in Ladakh, blocks the eastward moving moisture-laden wind system. Another contributing factor is regional pressure differences as a consequence of elevation (Example: Nubra Valley is lower in elevation than Leh Valley and tends to be greener) (Humbert-Droz and Dawa 2004; Shukla and Srivastava 2020).

Winter snow depth influences the vegetation growth of many land cover types (Peng et al. 2010; Dietz et al. 2012). Winter precipitation due to the Western Disturbances plays a vital role here, especially in the Zangskar and Suru regions, where the valley floors have more greenery due to the thick snow cover in the winter.

The regional vegetation at an elevation between 4300 and 5500 m a.s.l. across the Ladakh landscape is classified as desert-steppe and comprises around 40% of the area. The total vegetation is less than 5%, from alpine meadows on the north side of the Himalayan crest to 20% steppe vegetation to the north and east, with scrubland along river courses and 10% of subnival cushion vegetation (Demske et al. 2009). The region is devoid of trees due to the dry climate, except for isolated patches of juniper and birch in some valleys, and mostly cultivated varieties of poplar and willow along the major water courses (Fox et al. 1994; Hartmann 2009; Nüsser 2012; Doležal et al. 2018).

Ladakhi flora has considerable biological interest in view of its diversity of origins and important proportion of endemic species, depicting a high adaptability to extreme climatic conditions and biotic pressure (Chundawat and Rawat 1994).

The study region harbours a large number of plants of special conservation, economic and ethnobotanical significance: These include wild forage species and relatives of cultivated plants like onion and garlic (*Allium*), barley (*Hordeum vulgare*), gooseberry (*Ribes*), rhubarb (*Rheum*), wild legumes (*Cicer, Trigonella*), as well as a variety of aromatic and medicinal plants adapted to cold desert conditions (*Artemisia, Delphinium, Physochlaina, Tanacetum, Waldheimia*, etc.), occurring also in the peripheral areas to the west and south of Ladakh (Humbert-Droz and Dawa 2004; Doležal et al. 2018).

Several main areas of growth of medicinal plants in Ladakh have been identified (e.g., Khardong La, Taklang La, Padum region, Sapi region, Suru Valley, etc.). Non-medicinal plants including several tree and bush species have come under threat due to overcollection, such as junipers (*Juniperus*), which are mainly harvested for religious ceremonies and fuel, as well as birch (*Betula*), that have disappeared from much of their range in Ladakh (e.g., Nubra, Khaltse and the Nyemo-Zangskar highway area around Chiling) (Humbert-Droz and Dawa 2004; Doležal et al. 2018).

Fauna

Ladakhi fauna exhibits interesting characteristics that evolved as a result of adaptations to the extreme climatic conditions of the region, such as seasonal migration and hibernation, and amongst mammals, e.g., thick fur growth, thick and bushy tails, large nasal cavities, etc. (Chundawat and Rawat 1994).

The Blue Sheep or Bharal (local name: "Napo") and the Asiatic or Siberian Ibex ("Skin") are the only wild ungulates abundant in the region. The Tibetan Gazelle ("Gowa"), on verge of extinction, inhabits south-eastern Ladakh, the wild Yak ("Dong"), and Tibetan Antelope ("Tsos") inhabit in very limited numbers the eastern region of Chang Chenmo and the Aksai Chin region further north. The Tibetan Argali ("Nyan") is found in fragmented scarce numbers in the Changthang region. The Ladakh Urial ("Shapo"), another threatened wild sheep, is restricted to the lower slopes and banks of the main river valleys of the Shayok, Siachan, and Indus. Since their preferred habitat is heavily used for human settlements and agricultural purposes, it has led to its vulnerable status. The Tibetan wild ass ("Kyang") is another endangered ungulate, found at the flat bottom valleys and high plateaus of the Changthang region (Fox et al. 1991; Chundawat and Rawat 1994; Bhatnagar and Wangchuk 2001; Pfister 2004).

Amongst the carnivores the snow leopard ("Shan") is considered to be most under threat. It features a sparse but continuous distribution throughout the region. This elusive predator prefers the rugged and rocky terrain of central Ladakh. The Tibetan Wolf ("Shangku") is also classified as vulnerable, found in the open valleys and high plateaus of the region, less in western Ladakh. "This habitat preference together with its year round predation on livestock brings it into frequent conflict with local villagers and herders" (Mallon 1991; Pfister 2004).

A limited population of the threatened Brown Bear ("Denmo") exists in western Ladakh, predominantly in the Suru and Zangskar Valleys, while the Lynx ("I") managed to survive mainly in patches of undisturbed riverine scrubland. The confluence of the Shayok and Siachan rivers is an example of such a rare area. The Pallas's Cat ("Ribilik") is the rarest wild cat in the region, while other small carnivores such as the Red Fox ("Watse"), Mountain Weasel ("Lakimo"), and Stone Marten ("Kogar") are fairly abundant (Mallon 1991; Chundawat and Rawat 1994; Pfister 2004).

Birds of Ladakh can be distinguished in four main groups: The passage migrants form the first group that come in huge flocks displaying a great diversity in water birds (such as ducks, gulls, terns, etc.). They use the grass and wetlands of Ladakh for feeding and resting in autumn and early spring, as they migrate to and from their winter grounds on the subcontinent and beyond (Humbert-Droz and Dawa 2004; Pfister 2004).

The second group, summer visitors that migrate from the Indian subcontinent and Tibetan Plateau in spring, breed and spend the summer period in Ladakh. They consist mainly of water birds such as the exotic and vulnerable, Black-necked Crane ("Cha Thung Thung"), found nesting in the summer in parts of eastern Ladakh, near the wetlands of the Changthang region. The Bar-headed Goose ("Ngangpa") and

Ibisbill ("Tung Gyalmo") are other bird species that belong to this group (Humbert-Droz and Dawa 2004; Pfister 2004).

The third group, the resident birds, are well adapted to the Ladakhi extreme winter conditions. Examples are the Tibetan and Himalayan Snowcocks ("Ribzha"), Chukar Partridge ("Srakpa"), Oriental Skylark ("Zhingi Ama"), Tibetan Snowfinch ("Richi"), and some other birds that occur in more abundance, such as the Lammergeier ("Skyaklak"), Golden Eagle ("Laknak"), Rock Pigeon ("Phurgon"), Oriental Turtle Dove ("Kamru"), Raven ("Phorok"), Red-billed Chough ("Chungka"), and Black-billed Magpie ("Katang Putit") (Humbert-Droz and Dawa 2004; Pfister 2004).

The last group is very few winter visitors, such as the Northern Lapwing and the Dark-throated Thrush, that migrate in autumn and winter from their northern Palaearctic breeding grounds to the lower valleys of Ladakh (Humbert-Droz and Dawa 2004; Pfister 2004).

1.4 Structure of the Book

This book is comprised of five chapters. This introductory chapter is followed by the Materials and Methods sections of Chap. 2.

The Materials section deals primarily with the software, equipment, and satellite and field data used in this research. The Methods section is about the determination of a MODIS NDSI threshold value optimal for the study region Ladakh, employing the NDSI snow mapping algorithm (SNOMAP) method, and the subsequent validation against cloud-free Landsat images of four seasons, performing an accuracy assessment with Cohen's kappa coefficient method. It also deals with the development of algorithms to remove clouds obscuring the MODIS data, in order to generate low spatial but very high temporal resolution daily cloud-free snow maps and regional snow lines.

The spatio-temporal patterns and trends of the SCA and SCD were investigated over the whole study area and in seven delineated subregions, on the basis of mountain ranges and watershed areas. Moreover, in the subregions, based on topographical features such as elevation, slope, and aspect, the SCA was further analysed to detect and quantify trends.

The Methods section also contains the followed steps of a forecasting method for snow, on the basis of the MODIS snow data. And lastly the chapter is further enriched through on-ground measurements, with the installation of custom-made weather stations and in situ snow surveys. These efforts culminate in the generation of robust reference data, lending empirical heft to the broader research framework. The study also showcases the synergistic potential of modern technology by deploying terrestrial cameras across diverse topographies. These cameras provide high-resolution imagery that validates and augments the insights gleaned from remote sensing sources.

In Chap. 3 the results are presented. The procurement of de-clouded spatio-temporal images of the original MODIS snow data; the observed monthly, seasonal, and annual patterns and trends of the SCA on a regional and subregional level, and the pixel-wise assessment of the SCD on a seasonal and annual basis.

Thereupon, a further differentiation of the SCA in relation to elevation, slope, and aspect sheds light on many trend patterns in the seven subregions.

A statistical Autoregressive Integrated Moving Average (ARIMA) forecast model demonstrated credible scope for future snowfall predictions.

Field surveys yielded valuable data about the snow depth, SWE, and snow density, and temperature weather data could be correlated with the SCA and RSLE in the Gyamtsa and Khakshal locations. At the end of the chapter, Landsat NDSI snow lines established a good agreement with orthorectified photos of the Gyamtsa and Lungmar terrestrial cameras.

Chapter 4 consists of a discussion, and last Chap. 5 ends with conclusions and future outlook.

References

Ahmad I, Ahmad Z, Munir S, Shah SRA, Shabbir Y (2018) Geo-spatial dynamics of snowcover and hydro meteorological parameters of Astore basin, UIB, HKH Region, Pakistan. Arabian J Geosci 11:15. https://doi.org/10.1007/s12517-018-3734-9

Ajai, Nayak S, Kulkarni AV, Sharma A, Bahuguna I, Rathore B, Singh S, Patel J (2010) Snow and glacier studies. Space Applications Centre, ISRO, Ahmedabad, Gujarat

Baghel R, Nüsser M (2015) Securing the heights: the vertical dimension of the Siachen conflict between India and Pakistan in the Eastern Karakoram. Polit Geogr 48:24–36. https://doi.org/10.1016/j.polgeo.2015.05.001

Barnett TP, Adam JC, Lettenmaier DP (2005) Potential impacts of a warming climate on water availability in snow-dominated regions. Nature 438(7066):303–309

Bates B, Kundzewicz Z, Wu S, Palutikof J (2008) Climate change and water. Technical Paper of the Intergovernmental Panel on Climate Change 6:202. https://www.ipcc.ch/site/assets/uploads/2018/03/climate-change-water-en.pdf

Bhambri R, Bolch T, Kawishwar P, Dobhal D, Srivastava D, Pratap B (2013) Heterogeneity in glacier response in the upper Shyok valley, northeast Karakoram. Cryosphere 7(5):1385–1398. https://doi.org/10.5194/tc-7-1385-2013

Bhan S, Devrani A, Sinha V (2015) An analysis of monthly rainfall and the meteorological conditions associated with cloudburst over the dry region of Leh (Ladakh), India. Mausam 66(1):107–122. https://doi.org/10.54302/mausam.v66i1.371

Bhasin V (2012) Life on an edge among the Changpas of Changthang, Ladakh. J Biodivers 3(2):85–129. https://doi.org/10.1080/09766901.2012.11884738

Bhatnagar Y, Wangchuk R (2001) Status survey of large mammals in eastern Ladakh and Nubra. In: Conservation Biodiversity in the Trans-Himalaya: New Initiatives for Field Conservation in Ladakh, pp 108–125

Bhutiyani M (1999) Mass-balance studies on Siachen glacier in the Nubra valley, Karakoram Himalaya, India. J Glaciol 45(149):112–118. https://doi.org/10.3189/S0022143000003099

Bhutiyani M, Kale VS, Pawar N (2007) Long-term trends in maximum, minimum and mean annual air temperatures across the Northwestern Himalaya during the twentieth century. Climatic Change 85(1–2):159–177. https://doi.org/10.1007/s10584-006-9196-1

Bhutiyani M, Kale V, Pawar N (2010) Climate change and the precipitation variations in the northwestern Himalaya: 1866–2006. Int J Climatol: A J R Meteorol Soc 30(4):535–548. https://doi.org/10.1002/joc.1920

Bilal H, Chamhuri S, Mokhtar MB, Kanniah KD (2019) Recent snow cover variation in the Upper Indus Basin of Gilgit Baltistan, Hindukush Karakoram Himalaya. J Mountain Sci 16(2):296–308. https://doi.org/10.1007/s11629-018-5201-3

Brombierstäudl D, Schmidt S, Nüsser M (2021) Distribution and relevance of aufeis (icing) in the Upper Indus Basin. Sci Total Environ 780:146604. https://doi.org/10.1016/j.scitotenv.2021.146604

Brombierstäudl D, Schmidt S, Nüsser M (2022) Spatial and temporal dynamics of aufeis in the Tso Moriri basin, eastern Ladakh, India. Permafrost Periglacial Process 34:81–93. https://doi.org/10.1002/ppp.2173

Chevuturi A, Dimri A, Thayyen R (2018) Climate change over Leh (Ladakh), India. Theor Appl Climatol 131(1-2):531–545. https://doi.org/10.1007/s00704-016-1989-1

Chorol L, Gupta SK (2023) Evaluation of groundwater heavy metal pollution index through analytical hierarchy process and its health risk assessment via Monte Carlo simulation. Process Saf Environ Protection 170:855–864. https://doi.org/10.1016/j.psep.2022.12.063

Chundawat R, Rawat G (1994) Indian cold deserts: a status report on biodiversity. Wildlife Institute of India, Dehradun

Cogley G (2012) No ice lost in the Karakoram. Nat Geosci 5(5):305–306. https://doi.org/10.1038/ngeo1456

Dahri ZH, Ludwig F, Moors E, Ahmad B, Khan A, Kabat P (2016) An appraisal of precipitation distribution in the high-altitude catchments of the Indus basin. Sci Total Environ 548:289–306. https://doi.org/10.1016/j.scitotenv.2016.01.001

Dai L, Che T, Ding Y, Hao X (2017) Evaluation of snow cover and snow depth on the Qinghai-Tibetan Plateau derived from passive microwave remote sensing. Cryosphere 11(4):1933–1948. https://doi.org/10.5194/tc-11-1933-2017

Dame J (2018) Food security and translocal livelihoods in high mountains: evidence from Ladakh, India. Mount Res Dev 38(4):310–322. https://doi.org/10.1659/MRD-JOURNAL-D-18-00026.1

Dame J, Nüsser M (2008) Development perspectives in Ladakh, India. Geographische Rundschau International Edition 4(4):20–27

Dame J, Schmidt S, Müller J, Nüsser M (2019) Urbanisation and socio-ecological challenges in high mountain towns: insights from Leh (Ladakh), India. Landscape Urban Plan 189:189–199. https://doi.org/10.1016/j.landurbplan.2019.04.017

Dankers R, De Jong SM (2004) Monitoring snow-cover dynamics in Northern Fennoscandia with SPOT VEGETATION images. Int J Remote Sensing 25(15):2933–2949. https://doi.org/10.1080/01431160310001618374

Demske D, Tarasov PE, Wünnemann B, Riedel F (2009) Late glacial and Holocene vegetation, Indian monsoon and westerly circulation in the Trans-Himalaya recorded in the lacustrine pollen sequence from Tso Kar, Ladakh, NW India. Palaeogeogr Palaeoclimatol Palaeoecol 279(3–4):172–185. https://doi.org/10.1016/j.palaeo.2009.05.008

Dharpure JK, Patel A, Goswami A, Kulkarni AV, Snehmani (2020) Spatiotemporal snow cover characterization and its linkage with climate change over the Chenab river basin, western Himalayas. GIScience Remote Sensing 57(7):882–906. https://doi.org/10.1080/15481603.2020.1821150

Dharpure JK, Goswami A, Patel A, Kulkarni AV, Snehmani (2021) Assessment of snow cover variability and its sensitivity to hydrometeorological factors in the Karakoram and Himalayan region. Hydrol Sci J 66(15):2198–2215. https://doi.org/10.1080/02626667.2021.1985125

Dietz AJ, Kuenzer C, Gessner U, Dech S (2012) Remote sensing of snow—a review of available methods. Int J Remote Sensing 33(13):4094–4134. https://doi.org/10.1080/01431161.2011.640964

Dimri A, Chevuturi A (2014) Model sensitivity analysis study for western disturbances over the Himalayas. Meteorol Atmos Phys 123(3):155–180. https://doi.org/10.1007/s00703-013-0302-4

Doležal J, Dvorský M, Börner A, Wild J, Schweingruber FH (2018) Anatomy, age and ecology of high mountain plants in Ladakh, the Western Himalaya. Springer, New York. https://doi.org/10.1007/978-3-319-78699-5

Dong C, Menzel L (2016) Producing cloud-free MODIS snow cover products with conditional probability interpolation and meteorological data. Remote Sensing Environ 186:439–451. https://doi.org/10.1016/j.rse.2016.09.019

Downs RM, Day FA, Knox PL (1999) The national geographic desk reference. National Geographic Society, Washington, DC

Fox JL, Nurbu C, Chundawat RS (1991) The mountain ungulates of Ladakh, India. Biol Conserv 58(2):167–190. https://doi.org/10.1016/0006-3207(91)90118-S

Fox JL, Nurbu C, Bhatt S, Chandola A (1994) Wildlife conservation and land-use changes in the Transhimalayan region of Ladakh, India. Mount Res Dev 14(1):39–60. https://doi.org/10.2307/3673737

Gafurov A, Bárdossy A (2009a) Cloud removal methodology from MODIS snow cover product. Hydrol Earth Syst Sci 13(7):1361–1373. https://doi.org/10.5194/hess-13-1361-2009

Gafurov A, Bárdossy A (2009b) Snow cover data derived from MODIS for water balance applications. Hydrol Earth Syst Sci Discuss 6(1):791–841. http://doi.org/10.5194/hessd-6-791-2009

Goodison B, McKay D (1978) Canadian snowfall measurements: some implications for the collection and analysis of data from remote stations. In: Proceedings of the 46th Annual Western Snow Conference, Western Snow Conference, Oregon, Canada, pp 48–57. https://westernsnowconference.org/sites/westernsnowconference.org/PDFs/1978Goodison.pdf

Goodison B, Brown R, Crane R, Alley R, Bales R, Barber D, Barry R, Bentley C, Carrol T, Cline D, Duguay C (1999) Cryospheric systems. In: EOS Science Plan: The State of Science in the EOS Program, NASA/Goddard Space Flight Center, Greenbelt, Maryland, USA, pp 261–307. https://eospso.nasa.gov/sites/default/files/publications/SciencePlan.pdf

Gray DM, Male DH (1981) Handbook of snow: principles, processes, management & use. Pergamon Press, Oxford

Gusain H, Mishra V, Bhutiyani M (2014) Winter temperature and snowfall trends in the cryospheric region of north-west Himalaya. Mausam 65(3):425–432. https://doi.org/10.54302/mausam.v65i3.1053

Hall DK (1995) Satellite snow-cover mapping: a brief review. In: Proc. 1st Moderate Resolution Imag. Spectroradiom.(MODIS) Workshop Snow Ice, NASA/Goddard Space Flight Center, Greenbelt, Maryland, USA, pp 15–22

Hall DK, Riggs GA (2007) Accuracy assessment of the MODIS snow products. Hydrol Process 21(12):1534–1547. https://doi.org/10.1002/hyp.6715

Hartmann H (2009) A summarizing report on the phytosociological and floristical explorations (1976-1997) in Ladakh (India). Landquart VBA, Landquart

Hinkler J, Pedersen SB, Rasch M, Hansen BU (2002) Automatic snow cover monitoring at high temporal and spatial resolution, using images taken by a standard digital camera. Int J Remote Sensing 23(21):4669–4682. https://doi.org/10.1080/01431160110113881

Hodge HN (2013) Ancient futures: learning from Ladakh. Oxford University Press, Oxford

Humbert-Droz B, Dawa S (2004) Biodiversity of Ladakh: strategy and action plan. Sampark [in association with Ladakh Ecological Development Group], Calcutta

IMD (1999) Climatological Tables of Observatories in India: 1951–1980. India Meteorological Department (IMD), Director General of Meteorology, Pune

Immerzeel W, Pellicciotti F, Bierkens M (2013) Rising river flows throughout the twenty-first century in two Himalayan glacierized watersheds. Nat Geosci 6(9):742–745. https://doi.org/10.1038/ngeo1896

Immerzeel WW, Droogers P, De Jong S, Bierkens M (2009) Large-scale monitoring of snow cover and runoff simulation in Himalayan river basins using remote sensing. Remote Sensing Environ 113(1):40–49. https://doi.org/10.1016/j.rse.2008.08.010
Jain SK, Goswami A, Saraf AK (2008) Accuracy assessment of MODIS, NOAA and IRS data in snow cover mapping under Himalayan conditions. Int J Remote Sensing 29(20):5863–5878. https://doi.org/10.1080/01431160801908129
Jain SK, Goswami A, Saraf AK (2009) Role of elevation and aspect in snow distribution in Western Himalaya. Water Resources Manag 23(1):71–83. https://doi.org/10.1007/s11269-008-9265-5
Kamp U, Byrne M, Bolch T (2011) Glacier fluctuations between 1975 and 2008 in the Greater Himalaya Range of Zanskar, southern Ladakh. J Mount Sci 8(3):374–389. https://doi.org/10.1007/s11629-011-2007-9
Kendall MG (1948) Rank correlation methods. Charles Griffin and Co. Ltd., London
Khali H, Pandey P, Ali SN, Champatiray PK (2022) MODIS observed spatiotemporal variation of snow cover in Zanskar valley, North-Western Himalaya. Front Water 4:853146. https://doi.org/10.3389/frwa.2022.853146
Khan SI, Adams TE (2019) Indus river basin: water security and sustainability. Elsevier, Amsterdam. https://doi.org/10.1016/C2016-0-04119-0
Kulkarni AV, Singh S, Mathur P, Mishra V (2006) Algorithm to monitor snow cover using AWiFS data of RESOURCESAT-1 for the Himalayan region. Int J Remote Sensing 27(12):2449–2457. https://doi.org/10.1080/01431160500497820
Kumar M, Tiwari RK, Kumar K, Rautela KS (2022) Prediction of seasonal snow accumulation and depletion by SARIMA model using MODIS data. In: AQUA - Water Infrastructure, Ecosystems and Society p 15. http://doi.org/10.21203/rs.3.rs-2024923/v1
Majeed U, Rashid I, Sattar A, Allen S, Stoffel M, Nüsser M, Schmidt S (2021) Recession of Gya Glacier and the 2014 glacial lake outburst flood in the Trans-Himalayan region of Ladakh, India. Sci Total Environ 756:144008. https://doi.org/10.1016/j.scitotenv.2020.144008
Mallon DP (1991) Status and conservation of large mammals in Ladakh. Biol Conserv 56(1):101–119. https://doi.org/10.1016/0006-3207(91)90092-N
Mann HB (1945) Nonparametric tests against trend. Econom J Econom Soc 13(3):245–259. https://doi.org/10.2307/1907187
Misra A, Kumar A, Bhambri R, Haritashya UK, Verma A, Dobhal DP, Gupta AK, Gupta G, Upadhyay R (2020) Topographic and climatic influence on seasonal snow cover: Implications for the hydrology of ungauged Himalayan basins, India. J Hydrol 585:124716. https://doi.org/10.1016/j.jhydrol.2020.124716
Muhammad S, Thapa A (2021) Daily terra–aqua MODIS cloud-free snow and randolph glacier inventory 6.0 combined product (M* D10A1GL06) for high-mountain Asia between 2002 and 2019. Earth Syst Sci Data 13(2):767–776. https://doi.org/10.5194/essd-13-767-2021
Mukherji A, Molden D, Nepal S, Rasul G, Wagnon P (2015) Himalayan waters at the crossroads: Issues and challenges. Int J Water Resour Dev 31(2):151–160. https://doi.org/10.1080/07900627.2015.1040871
Müller J (2022) Urban mountain waterscapes in Leh, Indian trans-Himalaya: the transformation of hydro-social relations. Springer, New York. https://doi.org/10.1007/978-3-031-18249-5
Müller J, Dame J, Nüsser M (2020) Urban mountain waterscapes: the transformation of hydro-social relations in the trans-Himalayan Town Leh, Ladakh, India. Water 12(6):1698. https://doi.org/10.3390/w12061698
Norman R (2019) Rang-Skat: a Ladakhi-English dictionary. Edited by Henk Thoma. Melong Publications, Leh, Ladakh
Norphel C, Tashi P (2015) Snow water harvesting in the cold desert in Ladakh: an introduction to artificial glacier. In: Mountain Hazards and Disaster Risk Reduction, Springer, Tokyo, Japan, pp 199–210. https://doi.org/10.1007/978-4-431-55242-0_11
Nüsser M (2001) Understanding cultural landscape transformation: a re-photographic survey in Chitral, eastern Hindukush, Pakistan. Landsc Urban Plan 57(3-4):241–255, https://doi.org/10.1016/S0169-2046(01)00207-9

Nüsser M (2012) Umwelt und Entwicklung im Himalaya: Forschungsgeschichte und aktuelle Themen. Geographische Rundschau 64(4):4–9 & map supplement. https://katalog.ub.uni-heidelberg.de/titel/67519506

Nüsser M, Baghel R (2014) The emergence of the cryoscape: contested narratives of Himalayan Glacier dynamics and climate change. In: Schuler B (ed), Environmental and Climate Change in South and Southeast Asia, Brill, Leiden, pp 138–156. https://doi.org/10.1163/9789004273221_007

Nüsser M, Baghel R (2016) Local knowledge and global concerns: artificial glaciers as a focus of environmental knowledge and development interventions. In: Meusburger P, Freytag T, Suarsana L (eds) Ethnic and Cultural Dimensions of Knowledge, vol Knowledge and Space 8. Springer, Heidelberg, pp 191–209. https://doi.org/10.1007/978-3-319-21900-4_9

Nüsser M, Schmidt S (2021) Glacier changes on the Nanga Parbat 1856–2020: a multi-source retrospective analysis. Sci Total Environ 785(147321). https://doi.org/10.1016/j.scitotenv.2021.147321

Nüsser M, Schmidt S, Dame J (2012) Irrigation and development in the Upper Indus Basin: characteristics and recent changes of a socio-hydrological system in Central Ladakh, India. Mount Res Dev 32(1):51–61. https://doi.org/10.1659/MRD-JOURNAL-D-11-00091.1

Nüsser M, Dame J, Kraus B, Baghel R, Schmidt S (2019a) Socio-hydrology of "artificial glaciers" in Ladakh, India: assessing adaptive strategies in a changing cryosphere. Reg Environ Change 19:1327–1337. https://doi.org/10.1007/s10113-018-1372-0

Nüsser M, Dame J, Parveen S, Kraus B, Baghel R, Schmidt S (2019b) Cryosphere-fed irrigation networks in the Northwestern Himalaya: precarious livelihoods and adaptation strategies under the impact of climate change. Mount Res Dev 39(2):R1–R11. https://doi.org/10.1659/MRD-JOURNAL-D-18-00072.1

Parajka J, Blöschl G (2008) Spatio-temporal combination of MODIS images–potential for snow cover mapping. Water Resour Res 44(3):13. https://doi.org/10.1029/2007WR006204

Parajka J, Bezak N, Burkhart J, Holko L, Hundecha Y, Krajči P, Mangini W, Molnar P, Sensoy A, Riboust P, Rizzi J, Thirel G, Arheimer B (2017) Estimation of snow line elevation changes from MODIS snow cover data. In: EGU General Assembly Conference Abstracts, European Geosciences Union General Assembly 2017, Vienna, vol 19, p 14873. https://meetingorganizer.copernicus.org/EGU2017/EGU2017-14873.pdf

Passang S, Schmidt S, Nüsser M (2022) Topographical impact on snow cover distribution in the trans-Himalayan region of Ladakh, India. Geosciences 12(8):311. https://doi.org/10.3390/geosciences12080311

Peng S, Piao S, Ciais P, Fang J, Wang X (2010) Change in winter snow depth and its impacts on vegetation in China. Global Change Biol 16(11):3004–3013. https://doi.org/10.1111/j.1365-2486.2010.02210.x

Pfister O (2004) Birds and mammals of Ladakh. Oxford University Press, Oxford

Philip G, Mazari R (2000) Shrinking lake basins in the proximity of the Indus Suture Zone of northwestern Himalaya: a case study of Tso Kar and Startsapuk Tso, using IRS-1C data. Int J Remote Sensing 21(16):2973–2984. https://doi.org/10.1080/01431160050144901

Romshoo SA, Murtaza KO, Shah W, Ramzan T, Ameen U, Bhat MH (2022) Anthropogenic climate change drives melting of glaciers in the Himalaya. Environ Sci Pollut Res 29(35):52732–52751. https://doi.org/10.1007/s11356-022-19524-0

Saydi M, Ding Jl (2020) Impacts of topographic factors on regional snow cover characteristics. Water Sci Eng 13(3):171–180. https://doi.org/10.1016/j.wse.2020.09.002

Scherer D, Hall DK, Hochschild V, Konig M, Winther Jg, Duguay CR, Pivot F, Matzler C, Rau F, Seidel K (2005) Remote sensing of snow cover. Geophys Monogr 163:7. https://doi.org/10.1029/163GM03

Schmidt S (2009) Die reliefabhängige Schneedeckenverteilung im Hochgebirge. Ein multiskaler Methodenverbund am Beispiel des Lötschentals (Schweiz), Bonner Geographische Abhandlungen, vol 123. E. Ferger, Bergisch Gladbach. https://hdl.handle.net/20.500.11811/10554

Schmidt S, Nüsser M (2012) Changes of high altitude glaciers from 1969 to 2010 in the Trans-Himalayan Kang Yatze Massif, Ladakh, northwest India. Arctic Antarctic Alpine Res 44(1):107–121. https://doi.org/10.1657/1938-4246-44.1.107

Schmidt S, Nüsser M (2017) Changes of high altitude glaciers in the Trans-Himalaya of Ladakh over the past five decades (1969–2016). Geosciences 7(2):27. https://doi.org/10.3390/geosciences7020027

Schmidt S, Weber B, Winiger M (2009) Analyses of seasonal snow disappearance in an alpine valley from micro-to meso-scale (Loetschental, Switzerland). Hydrol Process 23(7):1041–1051. https://doi.org/10.1002/hyp.7205

Schmidt S, Nüsser M, Baghel R, Dame J (2020) Cryosphere hazards in Ladakh: The 2014 Gya glacial lake outburst flood and its implications for risk assessment. Nat Hazards 104:2071–2095. https://doi.org/10.1007/s11069-020-04262-8

Sen PK (1968) Estimates of the regression coefficient based on Kendall's tau. J Am Stat Assoc 63(324):1379–1389. https://doi.org/10.1080/01621459.1968.10480934

Shafiq MU, Bhat M, Rasool R, Ahmed P, Singh H, Hassan H (2016) Variability of precipitation regime in Ladakh region of India from 1901–2000. J Climatol Weather Forecasting 4(2):4. http://doi.org/10.4172/2332-2594.1000165

Shukla AN, Srivastava S (2020) Flora of Ladakh: an annotated inventory of flowering plants. In: Biodiversity of the Himalaya: Jammu and Kashmir State, Springer, New York, pp 673–730. https://doi.org/10.1007/978-981-32-9174-4_26

Singh S, Rathore B, Bahuguna I (2014) Snow cover variability in the Himalayan–Tibetan region. Int J Climatol 34(2):446–452. https://doi.org/10.1002/joc.3697

Soheb M, Ramanathan A, Angchuk T, Mandal A, Kumar N, Lotus S (2020) Mass-balance observation, reconstruction and sensitivity of Stok glacier, Ladakh region, India, between 1978 and 2019. J Glaciol 66(258):627–642. https://doi.org/10.1017/jog.2020.34

Soheb M, Ramanathan A, Bhardwaj A, Sam L (2022) Spatiotemporal quantification of key environmental changes in Stok and Kang Yatze regions of Ladakh Himalaya, India. Geocarto Int 37(26):11509–11533. https://doi.org/10.1080/10106049.2022.2060312

Sood V, Singh S, Taloor AK, Prashar S, Kaur R (2020) Monitoring and mapping of snow cover variability using topographically derived NDSI model over north Indian Himalayas during the period 2008–19. Appl Comput Geosci 8:100040. https://doi.org/10.1016/j.acags.2020.100040

Steiner L (2011) Reconstruction of glacier states from geo-referenced, historical postcards. Master's thesis, Institut für Kartografie und Geoinformation, Eidgenössische Technische Hochschule Zürich, Zürich. https://www.geometh-data.ethz.ch/downloads/Steiner_Masterarbeit.pdf

Stockdale CA (2017) A century of landscape change in the southern Rocky Mountains and foothills of Alberta: Using historical photography to quantify ecological change. PhD thesis, University of Alberta, Dept of Renewable Resources, Edmonton, Alberta. https://doi.org/10.7939/R3W37M744

Thackeray CW, Derksen C, Fletcher CG, Hall A (2019) Snow and climate: feedbacks, drivers, and indices of change. Current Climate Change Rep 5:322–333. https://doi.org/10.1007/s40641-019-00143-w

Thapa S, Li B, Fu D, Shi X, Tang B, Qi H, Wang K (2020) Trend analysis of climatic variables and their relation to snow cover and water availability in the Central Himalayas: a case study of Langtang Basin, Nepal. Theor Appl Climatol 140(3):891–903. https://doi.org/10.1007/s00704-020-03096-5

Thayyen RJ, Dimri A, Kumar P, Agnihotri G (2013) Study of cloudburst and flash floods around Leh, India, during August 4–6, 2010. Natural Hazards 65:2175–2204. https://doi.org/10.1007/s11069-012-0464-2

Tran H, Nguyen P, Ombadi M, Hsu Kl, Sorooshian S, Qing X (2019) A cloud-free MODIS snow cover dataset for the contiguous United States from 2000 to 2017. Sci Data 6:180300. https://doi.org/10.1038/sdata.2018.300

Van Andel TH (1985) New views on an old planet. Continental drift and the history of earth. Cambridge University Press, Cambridge

Wani JM, Thayyen RJ, Ojha CSP, Gruber S (2021) The surface energy balance in a cold and arid permafrost environment, Ladakh, Himalayas, India. Cryosphere 15(5):2273–2293. https://doi.org/10.5194/tc-15-2273-2021

Winiger M, Gumpert M, Yamout H (2005) Karakorum–Hindukush–western Himalaya: assessing high-altitude water resources. Hydrol Process 19(12):2329–2338. https://doi.org/10.1002/hyp.5887

Wylie T (1959) A standard system of Tibetan transcription. Harvard J Asiatic Stud 22:261–267. https://doi.org/10.2307/2718544

Yatoo MI, Singh M, Kubrevi SS, Shabir M, Dar R, Angmo K (2014) Safeguarding a national asset: a review on problems faced by Pashmina farmers in Changthang and their amelioration. Indian J Anim Sci 84:1251–1255. https://epubs.icar.org.in/index.php/IJAnS/issue/view/1269

Zhang H, Zhang F, Zhang G, Che T, Yan W, Ye M, Ma N (2019) Ground-based evaluation of MODIS snow cover product V6 across China: implications for the selection of NDSI threshold. Sci Total Environ 651:2712–2726. https://doi.org/10.1016/j.scitotenv.2018.10.128

Zhang Y, Yan S, Lu Y (2010) Snow cover monitoring using MODIS data in Liaoning Province, Northeastern China. Remote Sensing 2(3):777–793. https://doi.org/10.3390/rs2030777

Ziegler AD, Cantarero SI, Wasson RJ, Srivastava P, Spalzin S, Chow WT, Gillen J (2016) A clear and present danger: Ladakh's increasing vulnerability to flash floods and debris flows. Hydrol Process 30(22):4214–4223. https://doi.org/10.1002/hyp.10919

Chapter 2
Data and Methods for Snow Cover Detection

Abstract In this chapter, a multi-spatial, multi-temporal, multi-scale, and multi-source approach is conducted to investigate the snow cover distribution in Ladakh. The power of advanced remote sensing techniques is harnessed, notably leveraging the MODIS daily snow cover products. These tools facilitate the meticulous mapping of snow cover extent and duration, thereby encapsulating the temporal and spatial dimensions of this vigorous phenomenon. A pivotal aspect of the methodology involves the development of innovative algorithms to mitigate cloud cover interference within MODIS satellite imagery, thus enhancing the precision of long-term snow cover monitoring and forecasting. The empirical core of this research resides in the detailed examination of snow cover area (SCA), snow cover days (SCD), and the topographical impact on the SCA in seven delineated subregions based on mountain ranges and watersheds through an assortment of statistical approaches. These range from linear regression to the Mann-Kendall (MK) test with a two-variance correction, Sen's slope trend, and Pearson's r and Spearman's ρ correlation coefficient tests. In situ field survey data and meteorological data of custom-made weather stations are collected to facilitate the analysing process, and the installation of cameras provide high-resolution imagery which validates using the monoplotting method and augments the insights gleaned from remote sensing sources.

Keywords Snow · MODIS · DEM · NDSI · Cohen's kappa coefficient · Snowboard · Automatic weather station · Monophotogrammetry · Ladakh · India

2.1 Materials

Advancements in recent remote sensing techniques and satellite capabilities at a high temporal frequency and spatial resolution enable the monitoring of snow in data-sparse areas, such as rugged mountain terrains and cold regions like the Himalayas (Panda et al. 2015; Thenkabail 2015). At a regional scale one of the most important sensors, and widely used for the purpose of snow detection

S. Passang, *Snow Cover Distribution and Dynamics*, Advances in Asian Human-Environmental Research, https://doi.org/10.1007/978-3-031-57692-8_2

and monitoring, is the Moderate Resolution Imaging Spectroradiometer (MODIS), primarily because its data is readily and freely available worldwide on daily basis (Hall et al. 2001; Gafurov and Bárdossy 2009; Paudel and Andersen 2011; Gafurov et al. 2016). Snow mapping is obtained using a combination of visible and infrared bands of the electromagnetic spectrum, to discriminate between the reflectance of snow and clouds, impacts of vegetation and suitable surface heterogeneity. The MODIS snow cover product with 500 m spatial and daily temporal resolution covering the study area was obtained, spanning almost two decades from February 2000 until December 2016 (version 5) and from January 2017 until October 2019 (version 6).

The Normalised Difference Snow Index (NDSI) SNOMAP method was used as the MODIS snow mapping algorithm, with a threshold value calibrated for the four seasons in the study area (Hall et al. 1995, 2001, 2002; Gurung et al. 2011; Thenkabail 2015). Landsat 8 Operational Land Imager (LC8 OLI) snow cover maps on cloudless days with 30 m resolution were used to validate the MODIS snow cover data.

Several digital elevation models (DEMs) varying in scales were utilised: The Shuttle Radar Topography Mission (SRTM) DEM at a regional area scale for the reduction of clouds, to extract regional snow lines in order to derive the snow cover ratio, and to estimate the impact of local topography on snow distribution using MODIS snow data. Furthermore the High Mountain Asia (HMA) and TerraSAR add-on satellite for Digital Elevation Measurement (TanDEM-X) DEMs were used at a local scale to delineate snow lines from terrestrial imagery taken at four locations in Leh Valley, in order to validate the Landsat and MODIS snow cover data (Schmidt 2009; Schmidt et al. 2009; Steiner 2011; Stockdale et al. 2015; Stockdale 2017).

2.2 Data Sources

2.2.1 MODIS Data

MODIS is a land, ocean, and lower atmosphere observing instrument. It employs a cross-track scan mirror, collecting optics and a set of individual detector elements, to provide imagery of the Earth's surface and clouds in 36 spectral bands ranging in wavelength from 0.4 μm to 14.4 μm. This includes visible light, near-infrared, short, mid, and long wave infrared. The spatial resolution is 250 m in bands 1–2, 500 m in bands 3–7, and 1000 m in bands 8–36 (Hall et al. 1995; Riggs et al. 2006; Paudel and Andersen 2011; Riggs et al. 2016).

The first MODIS instrument was integrated on the National Aeronautics and Space Administration (NASA)'s Earth Observing System (EOS) Terra satellite (AM), launched on December 18, 1999. It commenced with observations in February 2000. A second MODIS instrument was integrated on the EOS Aqua satellite (PM), launched on May 4, 2002. Both were established for long-term observations of the Earth's environment (Paudel and Andersen 2011).

The data that the instruments provide is improving the understanding of global processes and helps accurately predict global changes. The key Earth's surface objectives of MODIS are to study global vegetation and land cover, global land surface change, vegetation properties, surface albedo, surface temperature, and snow and ice cover on a daily or nearly daily basis (Riggs et al. 2006, 2016).

The MODIS Terra and Aqua satellites orbit at an altitude of 705 km above the Earth's surface in near polar sun synchronous circular orbits. Terra (MOD10A1) has a descending node and the satellite crosses the study area at approx. 10:30 a.m. local time, Aqua (MYD10A1) has an ascending node and the satellite crosses the study area at approx. 1:30 p.m. (Riggs et al. 2006, 2016).

The snow cover data used to investigate the study region consists of daily snow cover maps (both MOD10A1 and MYD10A1) of collection 5, with a spatial resolution of 500 m. Included are snow cover, snow albedo, fractional snow cover, and Quality Assessment (QA) data in compressed Hierarchical Data Format-Earth Observing System (HDF-EOS) format, along with corresponding metadata. The MODIS Reprojection Tool (MRT) was used to execute the data conversion from Hierarchical Data Format (HDF) to GeoTIFF file format (MODIS Reprojection Tool 2011).

MODIS snow cover maps consist of six observed land cover characteristics: snow, land or no snow, lake, ocean, snow-covered water body (lake ice), cloud, no decision, night, saturated MODIS sensor detector and fill (no data expected for pixel) (Paudel and Andersen 2011). The maps encompass the period from 24 February 2000 until 2016. From 2017 onward only collection 6 are provided by the distributor, again with a spatial resolution of 500 m. For a list of the MODIS satellite data (downloaded from https://nsidc.org) used in the study see Table 2.1.

Due to the discontinuation of MODIS collection 5 and the unavailability of binary raster coded snow cover products in collection 6, an adjustment to set the NDSI threshold value calibration was necessary. This was done to validate the MODIS snow cover data, before algorithms were developed for the generation of cloudless daily MODIS snow cover maps of the study region (Gafurov and Bárdossy 2009; Paudel and Andersen 2011; Dong and Menzel 2016). The algorithm processes were primarily created and executed in the Graphical Modeler tool of the Geographical Information System QGIS, and supplementary scripting was done in the Python Console of QGIS and with R, a programming language and free software environment for statistical computing and graphics.

MODIS satellite snow cover data was preferred in the study region, as it has a high temporal resolution and is daily available on a global level. Moreover, the added advantage of the interval observing window between the two satellites of just 3 hours, for monitoring the snow cover and reducing cloud obscuring of the Earth's surface. This positively benefits in distinguishing between snow and thin cirrus clouds, albeit the data are of lower spatial resolution compared to, e.g., Landsat or Sentinel (Parajka and Blöschl 2008; Gafurov and Bárdossy 2009; Paudel and Andersen 2011).

Table 2.1 Satellite data and digital elevation models (DEMs)used in the study

Data type	Acquisition date	Product ID	Spatial resolution m×m
MODIS Snow Cover Daily L3 Global 500 m Grid v.5[a]	Oct. 2000–Dec. 2000	MOD10A1 (92 tiles)	500
	Jan. 2001–Dec. 2001	MOD10A1 (346 tiles)	500
	Jan. 2002–Dec. 2002	MOD10A1 (355 tiles)	500
	Jan. 2003–Dec. 2003	MOD10A1 (355 tiles)	500
	Jan. 2004–Dec. 2004	MOD10A1 (365 tiles)	500
	Jan. 2005–Dec. 2005	MOD10A1 (365 tiles)	500
	Jan. 2006–Dec. 2006	MOD10A1 (362 tiles)	500
	Jan. 2007–Dec. 2007	MOD10A1 (365 tiles)	500
	Jan. 2008–Dec. 2008	MOD10A1 (362 tiles)	500
	Jan. 2009–Dec. 2009	MOD10A1 (365 tiles)	500
	Jan. 2010–Dec. 2010	MOD10A1 (365 tiles)	500
	Jan. 2011–Dec. 2011	MOD10A1 (365 tiles)	500
	Jan. 2012–Dec. 2012	MOD10A1 (366 tiles)	500
	Jan. 2013–Dec. 2013	MOD10A1 (365 tiles)	500
	Jan. 2014–Dec. 2014	MOD10A1 (364 tiles)	500
	Jan. 2015–Dec. 2015	MOD10A1 (365 tiles)	500
	Jan. 2016–Dec. 2016	MOD10A1 (354 tiles)	500
MODIS Snow Cover Daily L3 Global 500 m Grid v.5[a]	Jul. 2002–Dec. 2002	MYD10A1 (170 tiles)	500
	Jan. 2003–Dec. 2003	MYD10A1 (355 tiles)	500
	Jan. 2004–Dec. 2004	MYD10A1 (365 tiles)	500
	Jan. 2005–Dec. 2005	MYD10A1 (365 tiles)	500
	Jan. 2006–Dec. 2006	MYD10A1 (362 tiles)	500
	Jan. 2007–Dec. 2007	MYD10A1 (365 tiles)	500
	Jan. 2008–Dec. 2008	MYD10A1 (362 tiles)	500
	Jan. 2009–Dec. 2009	MYD10A1 (365 tiles)	500
	Jan. 2010–Dec. 2010	MYD10A1 (365 tiles)	500
	Jan. 2011–Dec. 2011	MYD10A1 (365 tiles)	500
	Jan. 2012–Dec. 2012	MYD10A1 (366 tiles)	500
	Jan. 2013–Dec. 2013	MYD10A1 (365 tiles)	500
	Jan. 2014–Dec. 2014	MYD10A1 (364 tiles)	500
	Jan. 2015–Dec. 2015	MYD10A1 (365 tiles)	500
	Jan. 2016–Dec. 2016	MYD10A1 (354 tiles)	500
MODIS Snow Cover Daily L3 Global 500 m Grid v.6[a]	Jan. 2017–Dec. 2017	MOD10A1 (364 tiles)	500
	Jan. 2018–Dec. 2018	MOD10A1 (365 tiles)	500
	Jan. 2019–Sept. 2019	MOD10A1 (273 tiles)	500
MODIS Snow Cover Daily L3 Global 500 m Grid v.6[a]	Jan. 2017–Dec. 2017	MYD10A1 (365 tiles)	500
	Jan. 2018–Dec. 2018	MYD10A1 (365 tiles)	500
	Jan. 2019–Sept. 2019	MYD10A1 (273 tiles)	500

(continued)

Table 2.1 (continued)

Landsat 8[b]	21 Jan. 2016	LC08_L1TP_147036_20160121_20170405	30
	6 Feb. 2016	LC08_L1TP_147036_20160206_20170405	30
	9 Mar. 2016	LC08_L1TP_147036_20160309_20170328	30
	28 May 2016	LC08_L1TP_147036_20160528_20170324	30
	24 Feb. 2017	LC08_L1TP_147036_20170224_20170301	30
	24 Feb. 2017	LC08_L1TP_147037_20170224_20170301	30
	20 Sept. 2017	LC08_L1TP_147036_20170920_20170930	30
	20 Sept. 2017	LC08_L1TP_147037_20170920_20170930	30
	31 Mar. 2018	LC08_L1TP_147036_20180331_20180405	30
	31 Mar. 2018	LC08_L1TP_147037_20180331_20180405	30
	3 Jun. 2018	LC08_L1TP_147036_20180603_20180615	30
	26 Nov. 2018	LC08_L1TP_147036_20181126_20181210	30
	26 Nov. 2018	LC08_L1TP_147037_20181126_20181210	30
	19 Apr. 2019	LC08_L1TP_147036_20190419_20190423	30
NGA SRTM 1 Arcsecond Global[b]		n32_e075_1arc_v3	30
		n32_e076_1arc_v3	30
		n32_e077_1arc_v3	30
		n32_e078_1arc_v3	30
		n32_e079_1arc_v3	30
		n33_e075_1arc_v3	30
		n33_e076_1arc_v3	30
		n33_e077_1arc_v3	30
		n33_e078_1arc_v3	30
		n33_e079_1arc_v3	30
		n34_e075_1arc_v3	30
		n34_e076_1arc_v3	30
		n34_e077_1arc_v3	30
		n34_e078_1arc_v3	30
		n34_e079_1arc_v3	30
		n35_e075_1arc_v3	30
		n35_e076_1arc_v3	30
		n35_e077_1arc_v3	30
		n35_e078_1arc_v3	30
		n35_e079_1arc_v3	30
TanDEM-X[c]		TDM1_DEM__04_N33E077_DEM	12
		TDM1_DEM__04_N34E077_DEM	12

[a] Data downloaded from https://nsidc.org.
[b] Data downloaded from https://earthexplorer.usgs.gov
[c] Data downloaded from https://tandemx-science.dlr.de

Data Levels

Data products from MODIS instruments are processed at data levels ranging from Level 0 to Level 4. Level 0 data is raw instrument data. Level 1A data is reconstructed, unprocessed data at full resolution, that contains ancillary information such as calibration and georeferencing parameters on top of Level 0 data. Level 1B data is Level 1A data that has been processed to sensor units and is geometrically and radiometrically corrected. Level 2 data is processed data derived with averaged geophysical variables. Level 3 data is processed completely and at a high consistency. Level 4 data is the result of an algorithm analysis that uses lower range data (Riggs et al. 2006, 2016).

2.2.2 *Landsat Data*

In this research Landsat 8 OLI data is used to set the NDSI threshold with the highest Cohen's kappa coefficient against MODIS NDSI data, and for the comparison of oblique terrestrial images with both Landsat 8 and MODIS satellite images, using the WSL Monoplotting Tool (MPT) as an additional ground truthing method. For a list of the Landsat 8 satellite data (downloaded from https://earthexplorer.usgs.gov) used in the study see Table 2.1.

Landsat, an Earth observation satellite, delivers the longest continuous remote sensing images of the Earth's surface, to study over a period more than 50 years a global perspective on natural changes, and provide a close-up view of humanity's impacts on a wide range of topics such as ecosystems, urban growth, food, water, human health, energy, disasters, and climate change. It is a joint venture between NASA and the United States Geological Survey (USGS) that began in 1972 with the launch of Landsat 1.

Landsat 8 was launched on February 11, 2013. It carries an Operational Land Imagery (OLI) instrument and a Thermal Infrared Sensor (TIRS). The OLI captures data with radiometric precision over a 12-bit dynamic range, including nine spectral bands along with a panchromatic band. The TIRS includes two spectral bands (Wulder et al. 2012; Ihlen 2019; US Geological Survey 2021).

2.2.3 *Digital Elevation Model Data*

Available DEMs for the study area at a regional scale, such as those of the SRTM developed by the National Geospatial-Intelligence Agency (NGA) and the NASA Jet Propulsion Laboratory (JPL), the Global Digital Elevation Model by the Advanced Spaceborne Thermal Emission and Reflection Radiometer (ASTER), and the Global Digital Surface Model by the Advanced Land Observing Satellite

(ALOS), were thoroughly examined and compared with each other. Tables 2.1 and 2.2 list all the DEMs.

The NGA void filled SRTM 1 Arcsecond global data was utilised on the basis of its highest detail level, without many artefacts (irregular indents and bumps that are not naturally present). ASTER and ALOS data are both void filled but showed lots of artefacts, making them of a lower standard. The NASA JPL SRTM data, although also void filled, appeared generally less detailed than the NGA SRTM data.

HMA and TanDEM-X DEMs were used at a local scale.

SRTM

The SRTM launched aboard the Space Shuttle Endeavour on February 11, 2000 and landed again on February 22 at John F. Kennedy Space Center.

The SRTM used radar instruments to collect data for the most detailed near global topographic map of the Earth at the time. For this application radar is a much better tool than regular optical cameras, because it is able to operate both day and night and penetrate cloud cover.

Orbiting the Earth 16 times each day, over a period of 11 days the SRTM radar used interferometry to map over 80% of the Earth's land surface, i.e., more than 119 million km^2, between about 60° northern and 56° southern latitudes.

The radar images provide a representation of the Earth's surface viewed from two different positions, enabling researchers to derive a precise DEM of the entire Earth's surface. Operating the radar instruments on the shuttle at a nominal altitude of 233 km above the Earth made physical access to locations was no longer an issue, producing 3-dimensional (3D) maps of some of the world most inaccessible and unexplored areas. Information collected by the SRTM is being used in a wide range of applications to aid the activities of scientists, the military, commercial and civilian users, etc. (Jet Propulsion Laboratory NASA 2023).

In this work a total of 20 SRTM 1 Arcsecond global tiles (downloaded from https://earthexplorer.usgs.gov), with a spatial resolution of 30 m, were used to cover the study area (see also Table 2.1). Four tiles were found to have some voids (no data or null values):

- n32_e076_1arc_v3
- n33_e076_1arc_v3
- n34_e076_1arc_v3
- n35_e076_1arc_v3

To fill the voids, four NASA JPL SRTM tiles (downloaded from https://earthexplorer.usgs.gov), covering the same area at the same resolution as the four void-containing tiles, were used to patch the missing data. Elevation, aspect, and slope maps of the study area, along with regional zones, were also derived from the DEM (Paudel and Andersen 2011).

Table 2.2 Additional DEMs used in the DEM comparison

Data type	Product ID	Spatial resolution m×m
NASA JPL SRTM Version 3.0 Global 1 arc second (SRTMGL1)[a]	N32E075	30
	N32E076	30
	N32E077	30
	N32E078	30
	N32E079	30
	N33E075	30
	N33E076	30
	N33E077	30
	N33E078	30
	N33E079	30
	N34E075	30
	N34E076	30
	N34E077	30
	N34E078	30
	N34E079	30
	N35E075	30
	N35E076	30
	N35E077	30
	N35E078	30
	N35E079	30
ASTER Global Digital Elevation Model V003[b]	ASTGTMV003_N32E075_dem	30
	ASTGTMV003_N32E076_dem	30
	ASTGTMV003_N32E077_dem	30
	ASTGTMV003_N32E078_dem	30
	ASTGTMV003_N32E079_dem	30
	ASTGTMV003_N33E075_dem	30
	ASTGTMV003_N33E076_dem	30
	ASTGTMV003_N33E077_dem	30
	ASTGTMV003_N33E078_dem	30
	ASTGTMV003_N33E079_dem	30
	ASTGTMV003_N34E075_dem	30
	ASTGTMV003_N34E076_dem	30
	ASTGTMV003_N34E077_dem	30
	ASTGTMV003_N34E078_dem	30
	ASTGTMV003_N34E079_dem	30
	ASTGTMV003_N35E075_dem	30
	ASTGTMV003_N35E076_dem	30
	ASTGTMV003_N35E077_dem	30
	ASTGTMV003_N35E078_dem	30
	ASTGTMV003_N35E079_dem	30

(continued)

Table 2.2 (continued)

ALOS Global Digital Surface Model (AW3D30)[c]	ALPSMLC30_N032E075_DSM	30
	ALPSMLC30_N032E076_DSM	30
	ALPSMLC30_N032E077_DSM	30
	ALPSMLC30_N032E078_DSM	30
	ALPSMLC30_N032E079_DSM	30
	ALPSMLC30_N033E075_DSM	30
	ALPSMLC30_N033E076_DSM	30
	ALPSMLC30_N033E077_DSM	30
	ALPSMLC30_N033E078_DSM	30
	ALPSMLC30_N033E079_DSM	30
	ALPSMLC30_N034E075_DSM	30
	ALPSMLC30_N034E076_DSM	30
	ALPSMLC30_N034E077_DSM	30
	ALPSMLC30_N034E078_DSM	30
	ALPSMLC30_N034E079_DSM	30
	ALPSMLC30_N035E075_DSM	30
	ALPSMLC30_N035E076_DSM	30
	ALPSMLC30_N035E077_DSM	30
	ALPSMLC30_N035E078_DSM	30
	ALPSMLC30_N035E079_DSM	30

[a] Data downloaded from https://earthexplorer.usgs.gov
[b] Data downloaded from https://search.earthdata.nasa.gov
[c] Data downloaded from https://www.eorc.jaxa.jp

HMA

The HMA 8-meter resolution DEM data sets are derived from Very-High-Resolution (VHR) optical, along- and cross-track stereoscopic imagery from DigitalGlobe satellites, collected in the period between 28 January 2002 to 24 November 2016. DigitalGlobe is a commercial space imagery and geospatial content vendor based in the US. They also operate civilian remote sensing spacecraft. Three HMA 8-meter DEM data sets are downloadable since December 2017 at https://nsidc.org:

- High Mountain Asia 8-meter DEM Mosaics Derived from Optical Imagery, Version 1
- High Mountain Asia 8-meter DEMs Derived from Along-track Optical Imagery, Version 1
- High Mountain Asia 8-meter DEMs Derived from Cross-track Optical Imagery, Version 1

The data cover the region from Tien Shan and Hindu Kush to the Eastern Himalayas (National Snow & Ice Data Center, CIRES 2023).

In this research HMA 8-m DEM data was used in calculating camera viewsheds for the validation of the snow cover extent in the Gyamtsa, Khakshal, and South

Table 2.3 Additional data applied in the study

Data type	Product ID	Spatial resolution m×m
Land Water Mask Derived L3 Global 250 m[a]	MOD44W.A2015001.h24v05.006. 2018033162728	250
High Mountain Asia 8-meter DEM Mosaics Derived from Optical Imagery, v.1[b]	HMA_DEM8m_MOS_20170716 _tile-437	8
Randolph Glacier Inventory 6.0[c]	13_rgi60_CentralAsia 14_rgi60_SouthAsiaWest	

[a] Data downloaded from https://earthexplorer.usgs.gov
[b] Data downloaded from https://nsidc.org
[c] Data downloaded from https://glims.org

Phulu terrestrial camera locations, and the delineation of snow elevation transition lines in the Gyamtsa camera location, using the MPT as an another ground truthing method. The HMA data used in the study is listed in Table 2.3.

TanDEM-X

The TanDEM-X satellite was launched on June 21, 2010. The primary objective of the mission is to generate a global, consistent, timely, and high-precision DEM, with an accuracy corresponding to the High-Resolution Terrain Information-3 (HRTI-3) standard, as the basis for a wide range of scientific research, operational and commercial DEM products.

The secondary mission objectives are based on cross-track interferometry (for hydrology, geology, land environment, renewable and oceanography applications) and along-track interferometry (for moving target detection and traffic monitoring applications), as well as the development of new techniques (for super resolution, bi-static, moving target detection and polarimetric Synthetic Aperture Radar (SAR) applications) (German Aerospace Center (DLR) 2023).

In this study TanDEM-X 12 m data (downloaded from https://tandemx-science.dlr.de and listed in Table 2.1) found application due to missing data in the HMA 8-m DEM. It was used in calculating camera viewsheds for the validation of the snow cover extent and delineating snow elevation transitional lines in the Lungmar camera location, using the MPT as an another ground truthing method.

2.2.4 *Field Data*

In this research, the following field data have been used: Fresh snowfall samples collected on a snowboard in Khakshal village (period 2015–2019), in situ snow course surveys conducted at South Phulu (period 2015–2018), weather data gathered from the Gyamtsa AWS (period 2015–2017) and the Khakshal AWS (period 2016–2019). Also, terrestrial oblique images obtained from four camera stations find their application for analyses, correlation, and validation of coarser remote sensing snow data. One of these stations (South Phulu) also recorded the snow depth of an inserted graduated snow stake at the site.

2.3 Methods

2.3.1 NDSI Threshold Value Calibration

Snow has a high reflectivity in the visible range (green, MODIS band 4: 0.545–0.565 μm) and strong absorption in the short-wave infrared (SWIR) spectral range (SWIR 1, MODIS band 6: 1.628–1.652 μm). Clouds reflect strongly in both ranges. According to Gurung et al. (2011); Thenkabail (2015), the NDSI, based on an algorithm developed for MODIS snow mapping by Hall et al. (1995), is a measurement of the relative magnitude of the characteristic reflectance difference snow index between the visible wavelength (0.3–0.7 μm) and short-wave infrared wavelength (1.4 μm and longer) reflectance of snow (Riggs et al. 2006, 2016).

The NDSI is a spectral band ratio that takes advantage of the spectral differences of snow in SWIR and visible spectral bands, to identify snow versus other features in a scene. At visible wavelengths snow cover is just as bright as clouds and is therefore difficult to distinguish from cloud cover. However, at 1.6 μm snow cover absorbs sunlight and, therefore, appears much darker than clouds. This allows the effective discrimination between snow cover and clouds (Hall et al. 1995; Chokmani et al. 2010; Sibandze et al. 2014; Gaffey and Bhardwaj 2020). The snow identification using the NDSI is defined in Eq. 2.1:

$$\frac{Green - SWIR}{Green + SWIR} \tag{2.1}$$

Values of NDSI $\geq$ 0.40 typically indicate the presence of snow. It can however misclassify water bodies, having absorption characteristics in the near-infrared part of the spectrum. Here one needs to mask water bodies from the snow classification. However, to set an NDSI threshold value suitable for a region is not always simple and much information of secondary importance is needed to discriminate between snow-covered and snow-free areas. The threshold depends on the variation of many factors such as local terrain, season, angle of solar incidence, depth of snow on the surface, water content, vegetation, etc.

The NDSI is calculated for MODIS Terra using band 4 (green) and band 6 (SWIR 1), for MODIS Aqua using band 4 and band 7 (SWIR 2) in collection 5 data, but according to Riggs et al. (2016), in collection 6 the same snow detection algorithm can be run with both Terra and Aqua using band 6 data. The snow identification for

MODIS Terra and Aqua can therefore be defined in Eq. 2.2:

$$\frac{MODIS_{Green} - MODIS_{SWIR}}{MODIS_{Green} + MODIS_{SWIR}} \quad (2.2)$$

The global NDSI threshold value of the MODIS snow cover products distributed by the National Snow and Ice Data Center (NSIDC) is 0.40, with near-infrared (NIR) reflectance (MODIS band 2) greater than 0.11 and band 4 reflectance greater than 0.10, as noted by Riggs et al. (2006). This threshold was initially developed and validated only in the USA, Canada, Europe, and Russia (Chokmani et al. 2010).

Because of its complex terrain and shallow fragmented snow cover, there is not an established NDSI threshold value for mapping the snow cover in the region of Ladakh.

In this study eight Landsat 8 snow cover images on cloudless days of four different seasons in 2017 and 2018 were produced using the algorithm SNOMAP, for a comparison of the snow pixel congruence with sixteen cloud-free MODIS images of the same dates (Hall et al. 1995; Sibandze et al. 2014). According to Chokmani et al. (2010), the SNOMAP algorithm was intended to be an improvement of snow detection using NDSI. The Landsat obtained snow cover was classified into binary snow cover maps based on reflectance of band 3 (green) and band 6 (SWIR 1), with an NDSI more than or equal to 0.40, and band 5 (NIR) more than 0.11 (water mask) (see Fig. 2.1). The Landsat snow cover images, featuring a higher 30 m resolution, were resampled to 500 m to match and validate the 500 m resolution MODIS snow cover data. The approach was based on the Landsat relative ground truth method (Hall et al. 2001; Salomonson and Appel 2004; Wang 2017; Zhang et al. 2017; Tran et al. 2019).

An NDSI study area on the basis of two adjacent Landsat tiles, which predominantly covered most parts of the study area, was chosen to calculate an NDSI threshold value suitable for the MODIS snow cover maps for the region of Ladakh (see Fig. 2.2).

An accuracy assessment was performed to check the highest kappa coefficient against a range of different MODIS NDSI threshold value images, clipped to match the Landsat images' extent. Moreover, a MODIS water mask of 250 m resolution (see Table 2.3), resampled to 500 m, was also applied to the MODIS NDSI data. The highest obtained kappa coefficient value for each MODIS threshold value image was regarded to feature the appropriate NDSI threshold value for the MODIS snow cover.

Cohen's kappa coefficient (κ) represents the level of agreement between two data sets, considering the possibility of the agreement occurring by chance (Cohen 1960). The formula is given in Eq. 2.3:

$$\kappa = \frac{p_o - p_e}{1 - p_e} \quad (2.3)$$

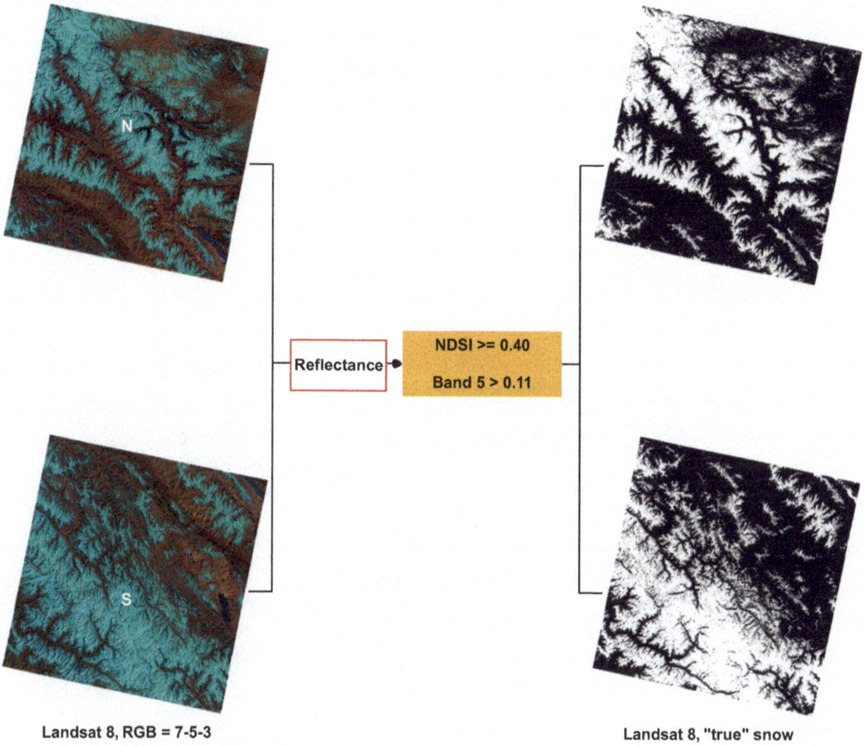

Fig. 2.1 NDSI SNOMAP method on the basis of Landsat 8 OLI images for relative ground truthing the snow cover of study region Ladakh (Images of 31 March 2018)

where p_o is the relative observed accuracy among rates and p_e the hypothetical probability of chance agreement.

2.3.2 Methods Applied for MODIS Cloud Removal

Cloud obscuration limits the dectection of snow coverage using optical remote sensing data, such as MODIS data. In the study region about 275 out of 365 days are sunny days. Based on the topography and climatic conditions of the area and on previous research reviews (Brombierstäudl et al. 2021), in this study an appropriate method for cloud filtering was developed. The MODIS data products were processed through three cloud filter algorithms, to facilitate the generation of daily cloudless MODIS snow cover maps.

Parajka and Blöschl (2008) first introduced three simple mapping methods termed as the combination of Terra and Aqua, the spatial filter, and the temporal filter that reduced cloud coverage over Austria. Later Gafurov and Bárdossy (2009)

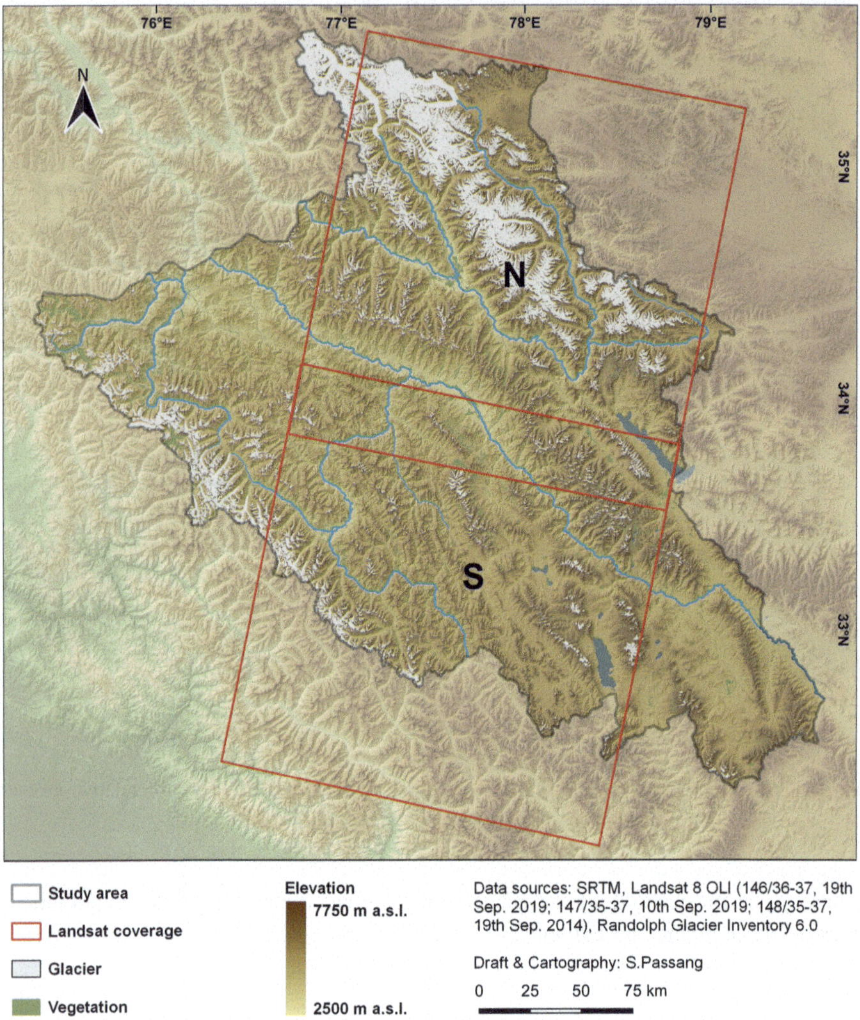

Fig. 2.2 NDSI study area on the basis of a Landsat 8 northern (N) and southern (S) tile. The smaller rectangle in the centre is where the tiles overlap

adopted some additional steps with modifications over the Kokcha Basin in the north-eastern part of Afghanistan, based on six steps that were developed to estimate the actual pixel cover for cloud-covered areas.

A regional Snow Line Mapping approach (SNOWL) was proposed by Parajka et al. (2010), for cloud reduction in the daily MODIS snow cover data set utilising elevation information. The SNOWL method is only effective in regions where cloud coverage for the entire region on a given day is smaller or equal to a cloud cover threshold. For this reason especially in regions where cloud cover remains large this approach cannot be applied. Dong and Menzel (2016), in south-west Germany, after utilising the first three simple methods proposed by Parajka and Blöschl (2008), utilised ground-based snow station information (meteorological interpolation) instead of SNOWL, to reduce the cloud cover from the daily MODIS snow cover data. The meteorological interpolation approach proved very effective in regions with dense snow station networks. It requires manual operation to set thresholds based on the location and altitude of the individual snow stations, for a more accurate prediction of the snow cover.

Krajčí et al. (2014) proposed a methodology to find elevation information based on the Regional Snow Line Elevation (RSLE), in which some snow pixels below, and land pixels above the RSLE are minimised. This method is however not very effective in a region with a shallow and patchy snow cover. Recently the same approach was used by Tran et al. (2019) in the USA, to mitigate clouds from the MODIS snow data.

The SNOWL method on the contrary shows more accurate results in an area with shallow and patchy snow. The reclassification of cloud pixels is based on the regional mean elevation of all snow and land pixels, respectively. Cloud pixels above the RSLE are assigned as snow, below the Regional Land Line Elevation (RLLE) as land, and between the RSLE and RLLE as mixed pixels. Therefore, in this research the SNOWL approach was deployed to minimise the cloud coverage and maximise the MODIS snow cover mapping accuracy.

In the study, suitable algorithms were developed to reduce the cloud obscuration of MODIS satellite images, adapted to the region's topography and climatic conditions, for an accurate outcome of the snow cover maps. Algorithms were based on Parajka and Blöschl (2008); Gafurov and Bárdossy (2009); Parajka et al. (2010); Krajčí et al. (2014); Dong and Menzel (2016). Each step, explained in the following sections, was performed in sequence and the output from each step was the input for the next step. The analysis was performed in QGIS, Python, and R. An overview of the sequence of the MODIS snow data processing, validated with Landsat 8 imagery and orthorectified photos from the camera stations is shown in flowchart format in Fig. 2.3.

Terra-Aqua Combination (T-A Combi) for Collection 5 Data

The combination of MOD10A1 and MYD10A1 snow cover area (SCA) images takes advantage of the 3-hour short interval time between two observations, and due to the constant movement of clouds during that time it is possible to reduce clouds, since the pixel coverage remains the same (Parajka and Blöschl 2008; Gafurov and Bárdossy 2009; Xie et al. 2009; Gao et al. 2010; Paudel and Andersen 2011; Krajčí et al. 2014; Dong 2016).

If a pixel was observed as cloud-covered by one satellite and cloud-free by another satellite, the resulting cloud-free pixel was assigned as the true pixel cover. This on the assumption that no snowmelt or snowfall occurred within this time shift.

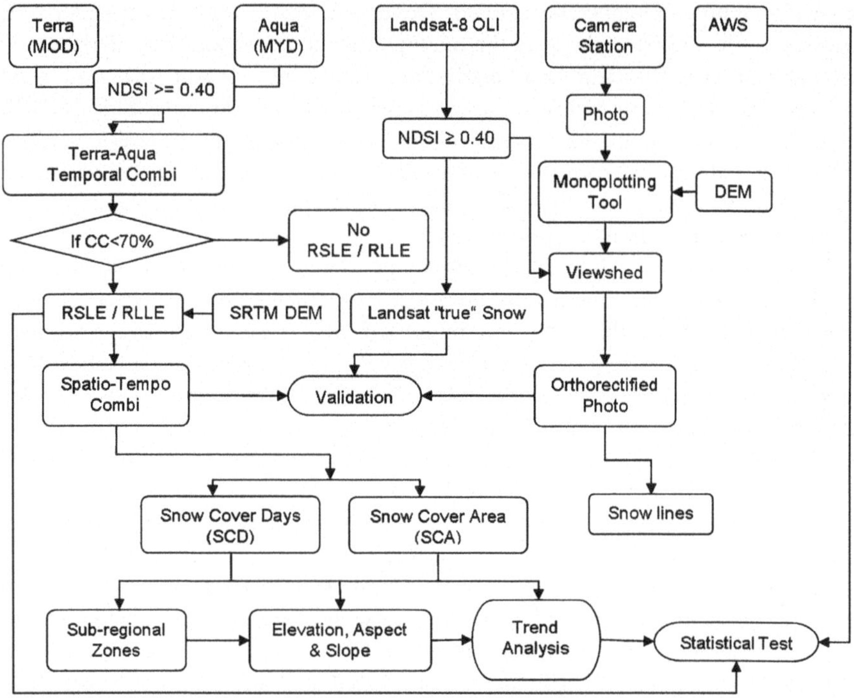

Fig. 2.3 Flow chart of MODIS snow data processing validated with Landsat 8 imagery and orthorectified photos from the camera stations

In this step first the pixel values of Terra and Aqua images were reclassified into three categories as snow (original snow and lake ice), land (no snow and lake classes) and cloud (all other missing/unusable classes) pixels, with unique values for Terra and Aqua, respectively. Subsequently the Terra and Aqua images were clipped to the study area of interest (AOI), and the ratios of the land, snow, and cloud cover calculated. These ratios indicate the proportions of, respectively, land, snow and cloud cover areas. They can be calculated using Eq. 2.4:

$$R_C = \frac{A_C}{A} \times 100\% = \frac{P^X}{P} \times 100\% \tag{2.4}$$

where R_C is the ratio of the land, snow, or cloud cover, A_C is the area of the land, snow, or cloud cover, A is the total area of the study area, P^X is the number of land, snow or cloud pixels, and P is the total number of pixels of the entire study area.

The clipped Terra and Aqua images were then combined and again reclassified as land, snow and clouds, and their cover ratios calculated. Result of this algorithm is a combination of collection 5 MODIS Terra and Aqua raster images of the same day, clipped to the study AOI, leading to cloud reduced images. The formula for the

Terra and Aqua combination is listed in Eq. 2.5:

$$P_{(x,y,t)} = \max\left(P^T_{(x,y,t)}, P^A_{(x,y,t)}\right) \tag{2.5}$$

where x is the longitude, y is the latitude, and t is a temporal (day) pixel P; P^T and P^A represent pixels from Terra and Aqua, respectively. In case either the Terra or Aqua image is missing, the sole image is used directly as the final output of the first step.

Terra-Aqua Combination (T-A Combi) for Collection 6 Data

This algorithm is an additional step to combine collection 6 MODIS Terra and Aqua raster images of the same day, in addition to the above-mentioned collection 5 suite, to continue the spatial-temporal MODIS snow cover analysis from 2017 onward.

In a preliminary step, collection 6 MODIS Terra and Aqua NDSI snow cover products were scaled by a factor of 0.0001, and the NDSI threshold value applicable to the study region had to be calibrated if necessary. Since the relevant average NDSI threshold value calculated from 16 different Terra and Aqua tiles over four seasons for the study area proved 0.40 (see Sect. 2.3.1), no changes were made to the original NDSI threshold value of 0.40 for snow. So if the NDSI $\geq$ 0.40, pixels are assigned as snow, the rest is no snow (Zhang et al. 2019).

To separate water bodies from snow, water masks were calculated from the water values of both Terra and Aqua NDSI snow cover products, and applied to the NDSI $\geq$ 0.40 images. Cloud masks were derived from the cloud values of both Terra and Aqua NDSI snow cover products and applied to the NDSI $\geq$ 0.40 images as well.

The pixel values of the Terra and Aqua images were then reclassified into three categories as land, snow and clouds pixels, with unique values for Terra and Aqua, respectively. Subsequently the same method for the combination of Terra and Aqua images for collection 5 data was used to reduce clouds from the data, and the ratios of land, snow, and cloud cover were calculated (using again Eq. 2.5).

MODIS snow cover collection 5 data was the first choice in this study, and the methods for cloud reduction had already been evaluated and analysed before the distributor stopped providing the collection in 2017. Terra and Aqua raster images of collection 5 contain already predefined NDSI threshold values, with pixel values pre-classified into distinct values for land, snow, water, ice, clouds, etc. Terra and Aqua NDSI raster images of collection 6, however, are rather based on fractional snow or no snow pixels, without fixed NDSI and pixel values. Therefore the above-mentioned water and cloud masks had to be applied and pixel values reclassified into three categories as land, snow, and clouds. With these few changes MODIS collection 6 could be successfully employed in this cloud filtering step (Riggs et al. 2016; Running et al. 2017).

Result of this algorithm is a combination of MODIS Terra and Aqua raster images of the same day, clipped to the study AOI, leading to cloud reduced images.

Temporal Combination (Tempo Combi)

The temporal combination algorithm step employs three observation windows. It combines the day before and the day after with the actual day of observation, on the assumption that both the previous and the next day of the cloud pixels have the same surface condition of either snow or no snow. So the temporal variation of 3 days was combined into a single day and then again reclassified into the three categories land, snow and clouds, with their cover ratios calculated (Parajka and Blöschl 2008; Gafurov and Bárdossy 2009; Gao et al. 2010; Tran et al. 2019). Results of this algorithm are the combination of collection 5 and/or collection 6 MODIS Terra and Aqua raster images over 3 days, leading to images with increased cloud reduction. The formula is given in Eq. 2.6:

$$P_{(x,y,t)} = 1,\ \text{if}\ (P_{(x,y,t-1)} = 1 \text{ and } P_{(x,y,t+1)} = 1) \tag{2.6}$$

where x is the longitude, y is the latitude, and t is a temporal (day) pixel P; $t - 1$ and $t + 1$ represent the preceding and succeeding days, respectively. 1 corresponds to snow cover and 0 to land cover.

Spatio-Temporal Combination (Spatio-Tempo Combi)

1. Snow transitional elevation line:
   The delineation of the RSLE and RLLE is the average (arithmetic mean) of elevations of all observed snow or land pixels of a temporal combination MODIS raster, being equal to the sum of the elevation values of each and every observed snow or land pixel, divided by the total number of snow or land pixels (Gafurov and Bárdossy 2009; Parajka et al. 2010; Paudel and Andersen 2011; Krajčí et al. 2014; Hauksson 2017; Tran et al. 2019). This can be shown in Eqs. 2.7 for RSLE and 2.8 for RLLE:

$$RSLE = \frac{1}{n}\sum_{i=1}^{n}\left(P^{S}_{(x,y,z)}\right)_i \tag{2.7}$$

$$RLLE = \frac{1}{n}\sum_{i=1}^{n}\left(P^{L}_{(x,y,z)}\right)_i \tag{2.8}$$

   where x is the longitude, y is the latitude, and z is the elevation of each snow pixel (P^S) or land pixel (P^L) in the list of elevations being averaged; and n is the total number of snow, or land pixels.
   The RSLE and RLLE form an intermediate step in reclassifying the pixel values of the remaining clouds as true pixels, assigned to either snow, land or mixed pixels, performed in the next step, the algorithm for the spatio-temporal

combination of images. The mean values of the RSLE and RLLE were extracted from the 500 m resampled SRTM DEM (Paudel and Andersen 2011).

The criteria of discarding images that have more than 75% cloud pixels was applied before processing the snow transitional elevation line algorithm (Paudel and Andersen 2011).

2. Reclassification of cloud pixels:

   In this step the pixel values of remaining clouds of the temporal combination algorithm were reclassified based on the RSLE and RLLE values of the snow transitional elevation line algorithm. This was performed in such a way that the remaining clouds were converted into land if the cloud pixels were below the mean RLLE and into snow if the cloud pixels were above the mean RSLE.

   The remaining cloud pixel values between the RSLE and RLLE were assigned as mixed pixels (patches of snow and land). Subsequently the snow, land, and mixed cover ratios were calculated. The mixed cover ratio indicates the proportion of the mixed cover area and can also be calculated with Eq. 2.4, where R_C then stands for the ratio of the mixed cover, A_C is the area of the mixed cover, and P^X is the number of mixed pixels (Parajka and Blöschl 2008; Paudel and Andersen 2011). Result of this algorithm is a completely cloud-free temporal combination MODIS raster image, where the remaining clouds are spatially distributed as snow, land and mixed pixels. The formula is given in Eq. 2.9:

$$P^C_{(x,y,z)} = \begin{cases} P^S_{(x,y,z)} & \text{if } (P^C_{(x,y,z)} > RSLE) \\ P^L_{(x,y,z)} & \text{if } (P^C_{(x,y,z)} < RLLE) \\ P^M_{(x,y,z)} & \text{if } (P^C_{(x,y,z)} < RSLE \text{ and } P^C_{(x,y,z)} > RLLE) \end{cases} \tag{2.9}$$

   where x is the longitude, y is the latitude, and z is the elevation value of a cloud (P^C), snow (P^S), land (P^L) or mixed pixel (P^M).

Figure 2.4 is an example of a heavily clouded day in the study area. It shows the cloud pixels (in grey) of a MODIS temporal combination (Fig. 2.4a) before reclassification into a MODIS spatio-temporal combination (Fig. 2.4b). The cloud pixels above the RSLE get assigned as snow pixels (in cyan) and below the RLLE as land pixels (in green). The remaining cloud pixels between the RSLE and RLLE become mixed pixels (in yellow).

2.3.3 Impact of Topography on Snow Cover Distribution

To investigate the impact of topographic (elevation, slope and aspect) effects on the snow cover distribution in the region, resampled SRTM DEM 500 m raster data (to match the MODIS data raster image resolution) was used to extract the elevation, slope and aspect of the entire study region (Jain et al. 2009; Paudel and Andersen 2011; Saydi and Ding 2020). It is important to note that prior to the clipping to

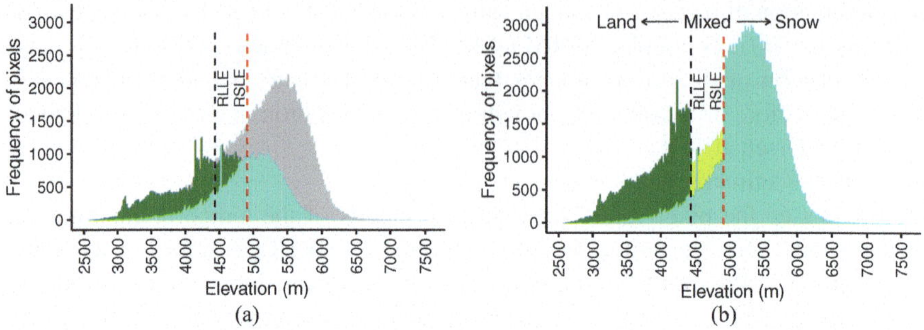

Fig. 2.4 Frequencies of pixels against elevation on 16 January 2017. (**a**) Snow pixels in cyan, land pixels in green and cloud pixels in grey (MODIS temporal combination of cloud reduction step 2) (**b**) Snow pixels in cyan, land pixels in green and mixed pixels in yellow (MODIS spatio-temporal combination of cloud reduction step 3). RLLE: 4433 m a.s.l. in a black dashed line and RSLE: 4919 m a.s.l. in a red dashed line

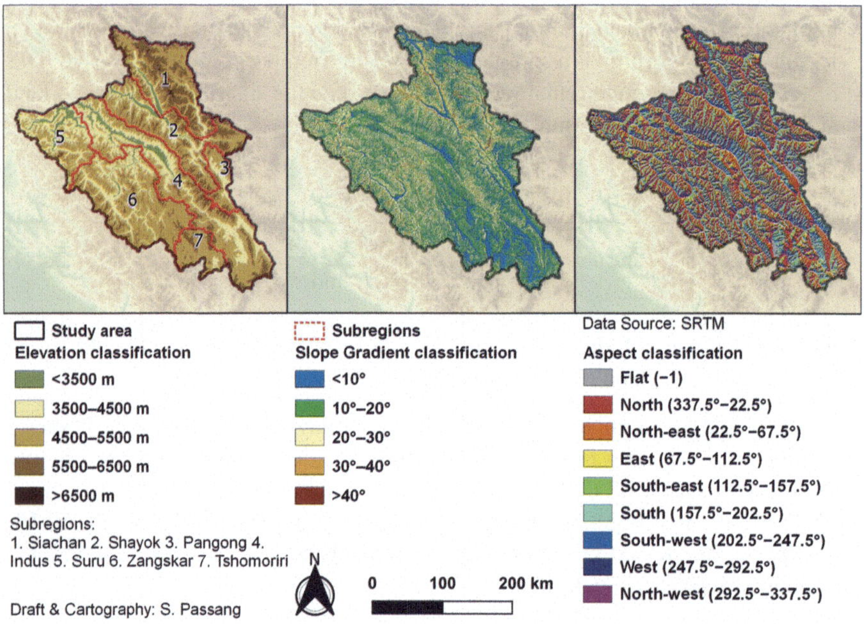

Fig. 2.5 Topographical parameters (spatial resolution of 500 m) of the study area: elevation (left), slope (middle), and aspect (right). Subregions are demarcated in red lines (left)

the study AOI, elevation, slope, and aspect were calculated based on the unclipped DEM, as to avoid miscalculations around the border edges.

The study area was delineated into seven subregions based on mountain ranges and watershed areas (see Fig. 2.5, left, for a visual display, and Table 2.4):

1. Siachan subregion: Siachan river originates from the Siachan glacier situated in the Karakoram range, north of the Nubra Valley. Some portion of the source of

Table 2.4 Details of the seven subregions in the study area

Subregion	Area km^2	Perimeter km	Max elevation m a.s.l.	Min elevation m a.s.l.	Mean elevation m a.s.l.
Siachan	10435	926	7740	3080	5300
Shayok	10567	1265	6770	2700	4980
Pangong	2543	514	6610	4220	5040
Indus	17112	1694	6660	2540	4700
Suru	6886	632	7110	2590	4390
Zangskar	14663	1135	6570	3090	4840
Tshomoriri	3322	494	6670	4310	5140

Shayok river was included in the Siachan zone due to its prominent snow cover throughout the season. Also commonly written Siachen.

Siachan (Tibetan: སེ་བ་ཅན་, Wylie: *se ba can*) is a compound of the two local terms "sia"—wild rose and "chan"—having (Norman 2019), referring to the many wild rose bushes in the area.

2. Shayok subregion: Shayok river is a major right bank tributary of the Indus river. It is flanked by the Karakoram and Ladakh range in Nubra Valley. Also commonly written Shyok.

   Shayok (Tibetan: ཤག་གཡོག་, Wylie: *shag g.yog*) is a contraction of the two local terms "shakma"—pebbles and "yok"—spread (Norman 2019), referring to the constantly fluctuating sedimentary structures that are left behind by the meandering river. A village along the river bears the same name.

3. Pangong subregion: Pangong lake (also called Pangong Tsho) is on the Changthang nomadic plateau region in eastern Ladakh. It is endorheic in nature and flanked by the Chang Chenmo and Pangong ranges. Approximately 40% of the length of the overall lake lies within Ladakh, 50% in Tibet and the rest is disputed but under the control of China.

   Pangong (Tibetan: སྤང་ཀོང་, Wylie: *spang kong*) is a compound of the two local terms "pang"—turf and "gong"—ditch (Norman 2019), referring to the large grassy valley surroundings of the lake.

4. Indus subregion: Indus river originates from the vicinity of Lake Manasarovar near Mount Kailash in Tibet. It is flanked by the Ladakh range and the Greater Himalayas. All the sub river basins within study area Ladakh become part of the Indus watershed basin.

   The Indus, locally known as Singge Khabaps (Tibetan: སེངྒེ་ཁ་བབས་, Wylie: *seng ge kha babs*), is a compound of the three local terms "sengge"—lion, "kha"—mouth and "babs"—descend, literally "(the river that) descends from the mouth of a lion" (Norman 2019).

5. Suru subregion: Suru river originates from the Panzella glacier. It is flanked by the western and northern boundaries of the Zangskar range and the Greater Himalayas. The origin of the name Suru is unknown.

6. Zangskar subregion: Zangskar river is a major left bank tributary of the Indus river. It is flanked by the Zangskar range and the Greater Himalayas, fed by several sub rivers. It is also commonly pronounced and written Zanskar.
   Zangskar (Tibetan: ཟངས་དཀར , Wylie: *zangs dkar*) is a contraction of the two local terms "zangs"—copper and "karpo"—white (Norman 2019), referring to the copper that used to be found in the area and a white alloy that was made from it.
7. Tshomoriri subregion: Tshomoriri lake is situated on the Changthang nomadic plateau region in south-eastern Ladakh. It is endorheic in nature and flanked by the Korzog range to the west and surrounded by some glaciers to the north, east and south.
   Tshomoriri (Tibetan: མཚོ་མོ་རི་རི , Wylie: *mtsho mo ri ri*) is based on a compound of the two local terms "tshomo"—lake and "ri"—mountain (Norman 2019).

The final MODIS snow data output (spatio-temporal combination) from the three cloud filtering algorithms was clipped to these seven subregions, to calculate the snow and mixed cover ratios per subregion. In this study mixed pixels are not used in any trend and correlation analysis.

To further examine in more detail the impact of elevation, slope and aspect on the snow cover distribution, the study area was divided into five elevation zones (<3500 m; 3500–4500 m; 4500–5500 m; 5500–6500 m; and >6500 m a.s.l.), five slope zones (<10°; 10°–20°; 20°–30°; 30°–40°; and >40°); and eight aspect zones (North (337.5°–22.5°); North-east (22.5°–67.5°); East (67.5°–112.5°); South-east (112.5°–157.5°); South (157.5°–202.5°); South-west (202.5°–247.5°); West (247.5°–292.5°); and North-west (292.5°–337.5°). See Fig. 2.5 for a visual presentation, and Table 2.5.

2.4 Snow Trend and Correlation Analysis

The monthly, seasonal, and annual MODIS snow cover data of the study area was examined on a regional and subregional level, in order to detect the occurrence of trends for 19 hydrological years (October 2000–September 2019). Seasons were categorised as autumn (September–November), winter (December–February), spring (March–May) and summer (June–August) of a hydrological year (October–September). A stepwise analysis was carried out in R and QGIS.

2.4.1 Descriptive Statistics and Linear Regression Model

Initially a descriptive statistics procedure was performed on a time series of the MODIS SCA and the RSLE, calculating minimum, maximum, mean, and standard deviation (SD). A time series is a sequence of observations over time, which are

Table 2.5 Spatial distribution of elevation, slope and aspect in the seven subregions of Ladakh

	Siachan		Shayok		Pangong		Indus		Suru		Zangskar		Tshomoriri	
	km^2	%	km^2	%	km^2	%	km^2	%	km^2	%	km^2	%	km^2	%
Elevation (m a.s.l.)														
<3500	263	3	512	5	–	–	1067	6	740	11	157	1	–	–
3500–4500	845	8	1808	17	546	21	5476	32	2946	42	3365	23	132	4
4500–5500	4898	47	5638	53	1383	54	8297	48	3122	45	9872	67	2441	73
5500–6500	4333	41	2622	25	614	24	2293	14	148	2	1298	9	749	23
>6500	115	1	1	0	–	–	1	0	4	0	0	0	1	0
Slope (°)														
<10	3216	31	2202	21	1090	43	5618	33	1118	16	3122	21	1633	49
10–20	3346	32	3975	38	1060	42	7616	44	2692	39	6075	41	1377	41
20–30	2595	25	3419	32	370	15	3565	21	2468	36	4575	31	303	9
30–40	1176	11	929	9	23	1	333	2	600	9	886	6	10	0
>40	119	1	54	1	–	–	2	0	37	1	34	0	–	–
Aspect														
North	1232	12	1285	12	231	9	2076	12	959	14	1593	11	320	10
North-east	1560	15	1473	14	445	17	2566	15	865	13	2177	15	427	13
East	1348	13	1556	15	369	14	2217	13	865	13	2048	14	445	13
South-east	1247	12	1188	11	295	12	1942	11	785	11	1639	11	328	10
South	1410	13	1112	11	224	9	1890	11	799	12	1533	10	371	11
South-west	1490	14	1260	12	333	13	2374	14	820	12	1971	13	490	15
West	1134	11	1385	13	299	12	2240	13	877	13	1993	14	463	14
North-west	1034	10	1316	12	236	9	1834	11	945	14	1739	12	381	11

usually spaced at regular intervals of time. A difference is made between a univariate time series, which refers to a set of observations over time of a single variable, and a multivariate time series, which refers to a set of observations over time of several variables. The objective of a time series analysis is to apply/develop models which are able to describe the given time series with a fair amount of accuracy (Brockwell and Davis 1991; Harvey and Shephard 1993; Box et al. 2008; Cryer and Chan 2008; Everitt and Skrondal 2010; Chatfield and Xing 2019; Hyndman and Athanasopoulos 2021).

Trends in the time series were further explored following the simple linear regression model using the least squares approach (Yan and Su 2009).

2.4.2 Mann–Kendall and Sen's Slope Tests

Subsequently Mann–Kendall (MK) and Sen's slope estimator (SSE) tests were carried out at three statistical significance levels at the confidence intervals of 90% (p-value (probability-value) < 0.1), 95% (p-value < 0.05) and 99% (p-value < 0.01), in order to detect and quantify the intensity of the SCA trends across the study area for 19 hydrological years (2000–2019). A MK test is an advanced statistical method for time series trend analysis of non-parametric data (Mann 1945; Kendall 1948). The test is based on the null hypothesis (H_0) that data in a time series have no trend. In case it finds evidence that the null hypothesis is false however, it is in favour of the alternative hypothesis, indicating a trend. The MK statistic S is given in Eq. 2.10:

$$S = \sum_{i=1}^{n-1} \sum_{j=i+1}^{n} sgn(x_j - x_i) \tag{2.10}$$

where x_i and x_j are sequential data values, n is the length of the data series, and

$$sgn(x_j - x_i) = \begin{cases} 1 & \text{if } (x_j - x_i) > 0 \\ 0 & \text{if } (x_j - x_i) = 0 \\ -1 & \text{if } (x_j - x_i) < 0 \end{cases} \tag{2.11}$$

The variance of S can be computed as in Eq. 2.12:

$$Var(S) = \frac{n(n-1)(2n+5) - \sum_{i=1}^{m} t_i(t_i - 1)(2t_i + 5)}{18} \tag{2.12}$$

where m is the number of tied groups and t_i is the number of data values in the ith group.

The standard normal variate Z is given in Eq. 2.13:

$$Z = \begin{cases} \frac{S-1}{(Var(S))^{\frac{1}{2}}} & \text{if } S > 0 \\ 0 & \text{if } S = 0 \\ \frac{S+1}{(Var(S))^{\frac{1}{2}}} & \text{if } S < 0 \end{cases} \tag{2.13}$$

Since the standard MK test of a time series is influenced by autocorrelation, obtained results are not error-free and need to be inspected. For this the Autocorrelation Function (ACF) was utilised. ACF in statistics is based on the Pearson Correlation Coefficient (PCC), developed by Karl Pearson (1857–1936). Autocorrelation checks for the linear relationship or correlation between data points of a time series separated by various time periods or time lags. The closely related Partial Autocorrelation Function (PACF) was also used that gives the partial correlation of a stationary time series with its own lagged values, with regressed values of the time series at the shorter time lags. The main difference with ACF is its control of other lags. ACF and PACF form part of the Box and Jenkins approach to time series modelling (Box et al. 2008; Chatfield and Xing 2019). Given time series measurements x_1, x_2,... x_n at time t_1, t_2,... t_n, the lag k autocorrelation function is defined in Eq. 2.14:

$$r_k = \frac{\sum_{i=1}^{n-k}(x_i - \overline{x})(x_{i+k} - \overline{x})}{\sum_{i=1}^{n-k}(x_i - \overline{x})^2} \tag{2.14}$$

where n is the length of the data series.

In case of detected autocorrelation, modified MK tests were applied. A modified variance of S as suggested by Yue and Wang (2004) is given in Eq. 2.15:

$$Var(S)^* = Var(S)\frac{n}{n^*} \tag{2.15}$$

where n is the actual sample size of the data, n^* is the effective sample size, and n/n^* is the correction factor.

The variance correction factors given by Hamed and Rao (1998) and Yue and Wang (2004) are, respectively, denoted by $C1$ and $C2$ in below Eqs. 2.16 and 2.17:

$$C1 = 1 + \frac{2}{n(n-1)(n-2)} \times \sum_{k=1}^{n-1}(n-k)(n-k-1)(n-k-2)\rho_k^R \tag{2.16}$$

$$C2 = 1 + 2\sum_{k=1}^{n-1}\left(1 - \frac{k}{n}\right)\rho_k \tag{2.17}$$

where ρ_k and ρ_k^R are the lag-k serial correlation coefficient of data and ranks of data, respectively.

An MK test is applied to identify whether there is a trend in the time series. If a trend exists, the magnitude can be determined by another non-parametric test, the Theil-Sen slope or Sen's slope estimator (Sen 1968). A robust method against outliers, it estimates the slope of the trend, given in Eq. 2.18:

$$\beta = median\left(\frac{x_j - x_i}{t_j - t_i}\right) \tag{2.18}$$

where x_j and x_i are data values at time t_j and t_i, respectively.

On a pixel-wise basis the seasonal and annual mean MODIS snow cover day (SCD) percentages were calculated for the entire study region of Ladakh, and the individual subregions.

A detailed study was conducted to analyse the spatio-temporal patterns and impact of topography (elevation, slope, and aspect) on the SCA, using a linear regression model, MK and modified MK tests in case of autocorrelation, and Sen's slope test, in order to detect and quantify the magnitude of the trends across the seven subregions.

2.4.3 Correlation Methods

To analyse correlations between the MODIS SCA, RSLE and temperature variables of the Khakshal and Gyamtsa AWSs, Pearson's r and Spearman's ρ correlation coefficient methods were conducted at 95% (p-value < 0.05) and 99% (p-value < 0.01) confidence limits.

2.4.4 Forecasting of Snow

The MODIS snow cover data was further analysed in R, in order to make an adequate statistical prediction model, that could be used to forecast the trend of future snowfall in the study area. Even though 30 years of data is historically considered the meteorological standard for climate research, several studies found that this is not always the optimal for future predictions. For example, a more recent 5–10 year period can have as much projective value as a 30 year record (WMO 2018). In this snow forecast analysis the 19 consecutive hydrological years in the period of October 2000 till September 2019 were used.

Time Series Decomposition and Stationarity

The time series of the snow cover data was initially decomposed and checked for stationarity. A time series generally comprises of a trend component, remainder, or noise component and can also have a seasonal component, in the case of a seasonal time series. Decomposing a time series means separating the original time series into those components, where two different structures can be used:

1. Additive decomposition—If the seasonal variation is relatively constant over time, one can use the additive model for decomposing a time series.
2. Multiplicative decomposition—If the seasonal variation is increasing over time, one can use the multiplicative model for decomposing a time series.

The additive and multiplicative decomposition models are defined in Eqs. 2.19 and 2.20:

$$y_t = S_t + T_t + R_t \tag{2.19}$$

$$y_t = S_t \times T_t \times R_t \tag{2.20}$$

where y_t is the value of the data at time t; S_t is the seasonal component at time t; T_t is the trend component at time t, and R_t is the remainder component at time t (Cryer and Chan 2008; Everitt and Skrondal 2010; Chatfield and Xing 2019; Hyndman and Athanasopoulos 2021).

A stationary time series has either a constant mean or variance or both that are constant over time. This makes a stationary time series easier to predict. There are several statistical tests that can be used to check the stationarity of a given time series. In this research, in addition to the in previous Sect. 2.4.2 performed ACF and PACF assessments, unit root tests like the Augmented Dickey-Fuller (ADF) and Kwiatkowski-Phillips-Schmidt-Shin (KPSS) were conducted (Dickey and Fuller 1979; Brockwell and Davis 1991; Kwiatkowski et al. 1992; Harvey and Shephard 1993; Fuller 1995; Box et al. 2008; McMurry and Politis 2010; Hyndman 2015; Hyndman and Athanasopoulos 2021).

The unit root ADF test is an augmented version of the Dickey-Fuller test (Dickey and Fuller 1979), for larger and more intricate time series. The test is based on the null hypothesis (H_0) that data in a time series are non-stationary. In case it finds evidence that the null hypothesis is false, it is in favour of the alternative hypothesis (H_1), which means that data in a time series are stationary (Dickey and Fuller 1979; Fuller 1995). Its statistic is defined in Eq. 2.21:

$$ADF = \rho.hat / S.E(\rho.hat) \tag{2.21}$$

where $\rho.hat$ is the coefficient estimation and $S.E(\rho.hat)$ is its corresponding estimation of standard error for each type of linear model. A p-value is calculated by interpolating the test statistics from the corresponding critical value tables for each type of linear models with given sample size n = length(x) (Fuller 1995).

The KPSS test is also a unit root test, similar to the ADF test. The null hypothesis states however that data in a time series are stationary; if false the time series follows the alternative hypothesis and is determined non-stationary (Kwiatkowski et al. 1992). KPSS is based on linear regression and its statistic is defined in Eq. 2.22:

$$x_t = r_t + \beta_t + \epsilon_t \tag{2.22}$$

where r_t is the random walk at time t; β_t is the deterministic trend (trend-stationary) at time t; and ϵ_t is the stationary error at time t.

In case the time series is non-stationary, one way to make it stationary is the computation of the differences between consecutive observations. This process is known as differencing (Cryer and Chan 2008; Everitt and Skrondal 2010; Chatfield and Xing 2019; Hyndman and Athanasopoulos 2021).

Forecast Model and Validation

To be able to model a time series event as a function of its past values, patterns in these values are identified and projected into the future. In this research the popular Autoregressive Integrated Moving Average (ARIMA) model was chosen, to best fit the snow time series. ARIMA models are common in weather forecasting in different parts of the world (Karamouz and Zahraie 2004; Şensoy and Uysal 2012; Lorentzen 2014; Sarhadi et al. 2014; Akpanta et al. 2015; Valipour 2015; Karimi et al. 2016; Chawsheen and Broom 2017; Shivhare et al. 2019; Dimri et al. 2020; Haq et al. 2020; Coban et al. 2021; Hyndman and Athanasopoulos 2021; Kumar et al. 2022).

ARIMA uses a number of lagged values in a time series, wherein a weight is applied to each of the past values, and the weights can vary based on how recent they are. The model is a refinement of the Autoregressive Moving Average (ARMA) model, first described by Whittle (1954), and further popularised by Box et al. (2008). For the ARIMA model to adequately fit a time series it is a requisite that the time series is stationary. It has three parameters that are used to help model the time series: seasonality, trend, and remainder or noise. These parameters are, respectively, labelled p, d, and q in the following Eq. 2.23:

$$ARIMA(p, d, q) \tag{2.23}$$

where p is the parameter associated with the autoregressive aspect of the model, which incorporates past values. For example, forecasting that if it snowed a lot over the past few days, you can state that it is likely it will snow tomorrow as well; d is the parameter associated with the integrated part of the model, which effects the amount of differencing applied to a time series. An example of this is forecasting that the amount of snow tomorrow will be similar to the amount of snow today, if the daily amounts of snow have been similar over the past few days; q is the parameter associated with the moving average part of the model (Chatfield and Xing 2019; Hyndman and Athanasopoulos 2021).

When the time series in the analysis has a seasonal component, a Seasonal Autoregressive Integrated Moving Average (SARIMA) model can be used. It uses univariate data containing trends and seasonality. In this case there are another three parameters, labelled P, D, and Q, that describe the same associations as the above mentioned p, d, and q but correspond only to the seasonal component of the model (Chatfield and Xing 2019; Hyndman and Athanasopoulos 2021). The SARIMA model can be written as Eq. 2.24:

$$ARIMA(p, d, q)(P, D, Q)[S] \tag{2.24}$$

where parameter S are the period units.

Appropriate parameter coefficients of the ARIMA model for the time series were calculated and the best ARIMA configuration for the training data (the first 80% of the snow data) examined. The predicted values were then compared to the test values (the latter 20% of the data), in order to determine the model accuracy (Hyndman and Athanasopoulos 2021).

In order to validate the efficiency of the chosen time series model, a commonly used method is Akaike's Information Criterion (AIC), an estimator of the prediction error (Akaike 1973, 1974). It can be expressed as Eq. 2.25:

$$AIC = T \log\left(\frac{SSE}{T}\right) + 2(k + 2) \tag{2.25}$$

where T is the number of observations used for estimation; SSE is the sum of the squared errors; and k is the number of predictors in the model. A time series model with the lowest AIC value is often the appropriate choice. For smaller T values, AIC tends to select too many predictors, and therefore a bias-corrected version of AIC was developed, also called the Corrected Akaike's Information Criterion (AIC_c) (Cavanaugh 1997; Burnham and Anderson 1998). Its formula is given in Eq. 2.26:

$$AIC_c = AIC + \frac{2(k+2)(k+3)}{T-k-3} \tag{2.26}$$

As with AIC, the lowest AIC_c value again indicates the best model. In this study both AIC and AIC_c were compared to validate the time series model.

After fitting a forecast model, the "residuals" or "white noise" in a time series model are what is left. They are similar to the remainder component after decomposing a time series (see previous Sect. 2.4.4). To further check the reliability of the ARIMA forecasting model, tests on the residuals of the model were performed, resulting in a graphical output in the form of a time series, an ACF correlogram, and a p-values plot of the Ljung-Box test. The Ljung-Box test is an improved version of the Box-Pierce test. It checks a chosen forecasting model by testing if the autocorrelation in the time series is different from zero. However, instead of testing randomness at each distinct lag, it tests the "overall" randomness based on a number of lags, and is therefore also called a portmanteau test (Box and Pierce 1970; Ljung and Box 1978; Box et al. 2008; Cryer and Chan 2008; Everitt and Skrondal 2010; Chatfield and Xing 2019; Hyndman and Athanasopoulos 2021). Its statistic is defined in Eq. 2.27:

$$Q = n(n+2)\sum_{k=1}^{h}\frac{r_k^2}{n-k} \tag{2.27}$$

where h is the number of lags tested; n is the number of data value observations in the time series; and r_k is the autocorrelation at lag k.

Lastly, a final calibration of the performance of the model was carried out by forecasting the snow cover data for the years 2015–2019, and validating the results with the actual data using a linear regression analysis.

2.5 Field Survey and Outdoor Measurements

During field surveys a Snow Metrics (Snowmetrics 2016) snow avalanche stick was used for point measurements, a Snow Metrics snow tube to measure snow depth, a Snow Metrics snow spring balance to measure snow water equivalent (SWE), a Garmin GPSMAP 64st GPS receiver to create ground control points and trace snow course lines for snow surveying plots; and an audio recorder was operated to record snow depth and SWE information.

During the winter of 2015/16, manual snow sampling of at least 30 point observations with at least 1 SWE each was performed at three different snow surveying plots, at elevations of 4770 m, 4810 m and 4860 m a.s.l., respectively.

In the hydrological year 2016/17, a time-lapse camera in combination with a graduated snow stake was set up at an elevation of 4900 m a.s.l.

During the winters of 2016/17 and 2017/18, manual snow sampling of at least 30 point measurements and 1 SWE was conducted at snow plot 1 (4770 m a.s.l.).

Snowboard snow sampling was carried out in Khakshal village, about 2.5 km north of Leh town at an elevation of approx. 3660 m a.s.l., during all winter periods of 2015–2019, simultaneously manually measuring snow depth and SWE, and automatically measuring snow depth with a Raspberry Pi (RPi) Camera Module (Raspberry Pi Foundation 2016), at an elevation of 3660 m a.s.l.

Table 2.6 summarises the snow sampling equipment and devices used in the study.

2.5.1 *Snowboard Meteorology*

The most direct method for measuring the depth of freshly fallen snow is obtained with a graduated ruler inserted vertically into the snow (Gray and Male 1981). A snow tube and snow spring balance (weighing scale) were used for the sampling of, respectively, snow depth and SWE. According to Pomeroy and Gray (1995), SWE is related to the snow depth and snow density as in Eq. 2.28:

$$SWE = HS * \rho_S \tag{2.28}$$

where SWE, HS, and ρ_S are, respectively, the SWE (in cm), snow depth (in cm) and snow density (in g cm^{-3}) of a snow sample.

Table 2.6 Snow samplers and additional devices

Item	Specification
Snow avalanche stick	Point measurement, resolution 1 cm
Snow tube	Snow depth, resolution 1 mm
Snow spring balance	SWE, resolution 1 mm
Snowboard with ruler	Snow depth, resolution 1 mm
Snow stake	Snow depth, resolution 1 cm
Raspberry Pi Camera Module	Snowboard snow depth photography
Bushnell Trophy camera	Snow stake snow depth photography
Garmin GPSMAP 60st	Snow sample points & snow course tracks
Audio recorder	Snow depth and SWE information

(a) (b)

Fig. 2.6 Snowboard. (**a**) Snowboard dimensions. (**b**) Snowboard with inserted ruler

A custom-made snowboard was made out of a piece of water-resistant plywood, with the dimensions 45 cm length by 45 cm width and 2 cm thick. It was painted white in order to minimise absorption by sunlight and provided a reference point for snow measurement. The snowboard was set up in November 2015 in Khakshal village (Fig. 2.6).

The surface of the snowboard was kept free of snow before any snowfall, in order to ensure that old snow would not be sampled (Gray and Male 1981). In addition to the manual snow sampling method on the snowboard, an RPi Camera Module (described in details in Sect. 2.6.3) is recording the snowboard with inserted ruler, acquiring the snow depth at an interval of 15 minutes during each winter, starting in January 2016 till the present. However for this study the observations were limited till the end of the hydrological year, September 2019.

2.5.2 *In Situ Snow Survey Observations*

Snowfall in the study region is relatively light due to the lee of the mountains, although the region experiences long winters. It is devoid of major moisture and extremely low temperatures reduce the moisture holding capacity of the air to very low values. One of the most difficult tasks was to find possible surveying sites with suitable accessibility in winter, to conduct field survey observations with low risk.

Reconnaissance of Survey Sites

A reconnaissance of the Gyamtsa Valley was conducted on 18 November 2015, and of the Stok Valley on 7 December, as to get an idea regarding possible surveying sites. However, in Stok the snow line was located on a steep side ridge above 5200 m a.s.l. and in Gyamtsa the south-facing ridges were all still barren.

(a) (b) (c)

Fig. 2.7 Acquiring in situ snow depth and SWE measurements in South Phulu. (**a**) Point measurement with a snow avalanche stick. (**b**) Snow depth measurement with a snow tube. (**c**) SWE measurement with a snow spring balance

On 10 December 2015, just a day after some fresh snowfall, a visit to the Khardong La pass area was made. At a trial plot at South Phulu, at an elevation of 4860 m a.s.l., a snow course was measured, recording 79 random snow depth point measurements with the avalanche stick, and 1 snow depth and SWE sample with the snow tube and snow spring balance, respectively (see Fig. 2.7).

Although this first attempt was after very little snow accumulation, it proved already exhausting and hazardous to reach a plot suitable for measuring a snow course. One had to traverse a terrain of rigid boulders and be extra cautious not to exert oneself much in the thin atmospheric air pressure, when one's body is not adapted to such high elevations and conditions, especially in the blistering cold of the winter. It took an hour just to measure one plot with an assistant.

Snow Surveys in South Phulu

In South Phulu snow surveys were performed to acquire the snow depth and SWE at three different plots at different elevations (see Fig. 2.8). During the first hydrological year 2015/16 a minimum of 30 point measurements and 1 SWE at each plot were conducted, usually at the beginning and the end of each month in an interval of 3 months (December–February).

In situ observations of the snow course performed at plot 1 were at an elevation of 4770 m a.s.l. and located on a gradual slope of the west-sloping face. The area was mostly exposed to the sun.

The snow course made at plot 2 was situated at an elevation of 4810 m a.s.l. and located at an area where it remained in the shadow after 1 p.m., due to an east-facing ridge to the west.

Plot 3 was situated at an elevation of 4860 m a.s.l. It was located on a flat stream basin, sandwiched between two ridges on each side, where it remained mostly in the shadow compared to the other two plots.

For the following hydrological years 2016/17 and 2017/18 some changes were made, due to the relatively close proximity of all three snow plots and the dangerous access and time constraints of the two higher ones. A Bushnell Trophy time-lapse

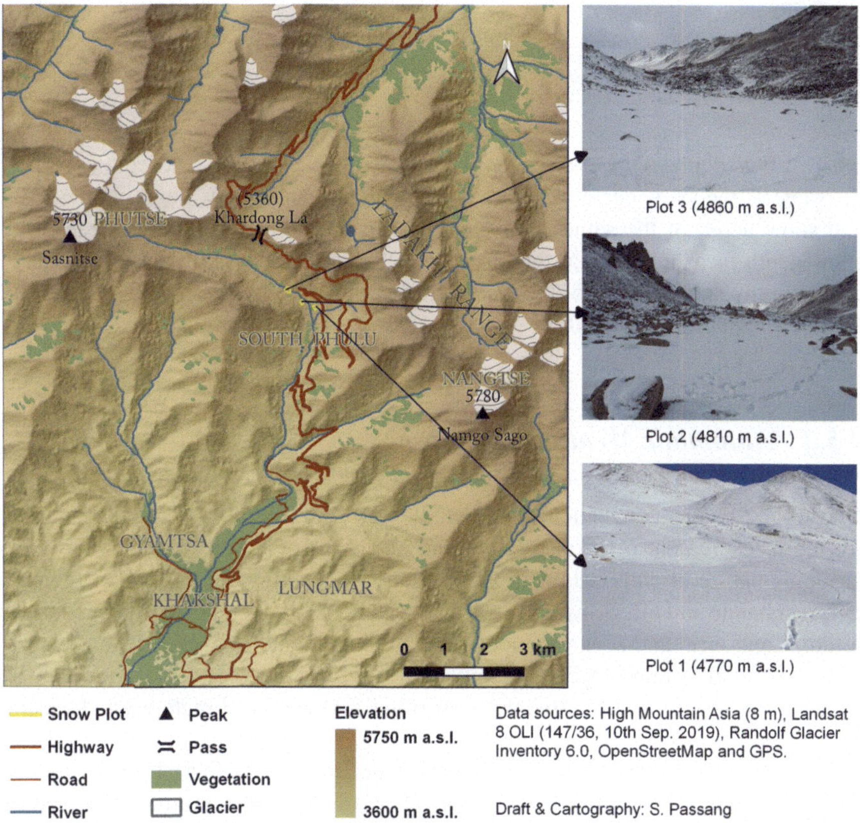

Fig. 2.8 Location of snow plots in South Phulu

camera (Bushnell Corporation 2016) and a graduated snow stake was set up at an elevation of 4900 m a.s.l., which proved to be a better alternative to traditional snow survey methods, in order to monitor snow cover information on a very high temporal resolution. With the exception of 7 January 2017, when all plots could be surveyed together for the last time, only lower plot 1 continued to be conducted during the next two hydrological winters, together with the time-lapse camera recording the snow stake at near-range and Phutse glacier in the background at far-range, for one complete hydrological year of 2016/17. Figure E shows the snow survey forms used during snow measurements.

Comparative to time-consuming snow depth and SWE measurements, the snow density tends to differ little over small regions (Martinec and Sevruk 1992; Schmidt 2009). Therefore, for a test region, where multiple snow depth, but only a single or few SWE measurements are taken, an overall SWE value can be calculated with

Eq. 2.29:

$$\overline{SWE} = \frac{SWE}{HS}\overline{HS} \tag{2.29}$$

where $\overline{SWE}$ denotes the average SWE of the snowpack at the test region; $\frac{SWE}{HS}$ is the same as the snow density, with SWE and HS, respectively, denoting the SWE and snow depth of the same snow sample; and $\overline{HS}$ is the arithmetic mean of all snow depth measurements at the test region. Since the snow plots in this research covered only small areas, for the calculation of the average SWE and snow density the above equation was used.

2.5.3 *Automatic Weather Stations*

One of the prime necessities in the study was the access to a weather station, for the measurement of local weather and environmental data. Understanding the changing nature of daily weather patterns helps uncovering the impacts of regional climate change and offers insights into how water resources could be managed in the future. Long-term and accurate monitoring of, e.g., the air temperature, barometric pressure, relative humidity, soil temperature, wind speed, and direction can be vital in developing an effective early warning system for climate related natural calamities.

Data from existing weather stations in the area is publicly limited, due to Ladakh's status as a sensitive border area. Alternatively several ready-made weather station solutions were researched on the internet; however, in India they proved relatively expensive and quality ones were not readily available. Therefore, the installation of a custom-made outdoor weather station became of primary importance, especially since it is multi-configurable and self programmable, in order to get accurate data.

Automatic weather stations (AWSs) that record local weather and environmental data using fully electronic solutions were developed and field-tested at sites in Gyamtsa and Khakshal. A similar station is described by Abdul-Niby et al. (2017).

2.6 Snow Line Analysis in Leh Valley

To measure snow cover patterns, time-lapse or time series photography involving oblique terrestrial images was used. Constant in situ measurements of snow are laborious and frequently impossible due to wind drift, snow redistribution, and the inaccessibility of the terrain. Therefore, time-lapse photography is a powerful tool for the collection of snow information at a relative low cost, especially at the small catchment scale and without the need for direct field access. Moreover, it provides a useful additional ground truthing method, as it is usually not obstructed by cloud cover, which is an advantage over optical remote sensing (Parajka et al. 2012; Dong and Menzel 2017).

Time-lapse photography has been widely used to monitor glacier and snow cover related processes in Europe and America, and to some extent also in China (Christiansen 2001; Floyd and Weiler 2008; Schmidt 2009; Schmidt et al. 2009; Farinotti et al. 2010; Parajka et al. 2012; Bernard et al. 2013; Garvelmann et al. 2013; van Herwijnen et al. 2013; Hedrick and Marshall 2014; Huintjes et al. 2015; Liu et al. 2015; Dong and Menzel 2017), but in the region of Ladakh this technique has never been used before.

While time-lapse photographs already provide a lot of informative details about snow cover conditions, since they are only in a "flat" two-dimensional (2D) spatially unreferenced format, their use in snow analysis is still limited. Advancements in computing power over the last two decades have enabled the development of techniques for the orthorectification of oblique terrestrial images. The principle of monophotogrammetry was employed to convert photographs from 2D to interpreted 3D or orthorectified, in order to produce spatially referenced data, for the exchange with a traditional Geographic Information System (GIS) (Tappeiner et al. 2001; Corripio 2002; Marco et al. 2018). In this section a new monoplotting software program called the MPT is described. The tool facilitates a fairly intuitive and user-friendly orthorectification of terrestrial images, permitting the rapid evaluation of large numbers of images. It has already been successfully used in many research papers, e.g., Steiner (2011); Stockdale et al. (2015); Stockdale (2017); Marco et al. (2018); Gabellieri and Watkins (2019).

2.6.1 The Monophotogrammetry Principle

Single (mono) oblique terrestrial images can be orthorectified using the technique of monophotogrammetry, also called monoplotting. This idea was already proposed in the early 1970s (Makarovic 1973). It was later also described in detail by, e.g., Eder and Aumann (1996); Strausz Jr (2001); Steiner (2011). Monophotogrammetry is in contrast with stereo-photogrammetry or stereo-pairs photography, a technique where a collection of overlapping oblique terrestrial images is used for the orthorectification procedure. Marco et al. (2018, p. 110) emphasised that "the main limitation in a monoplotting system is that in monophotogrammetry only points on the terrain (DEM surface) can be precisely located, whereas stereo-photogrammetry enables the calculation of the position of any common pixel in the stereo pair."

The monoplotting technique assumes that a camera, a point on a 2D photograph, and the 3D point in the real world can all be related to each other when positioned in a straight line. Figure 2.9 visually shows this relationship, which is described by

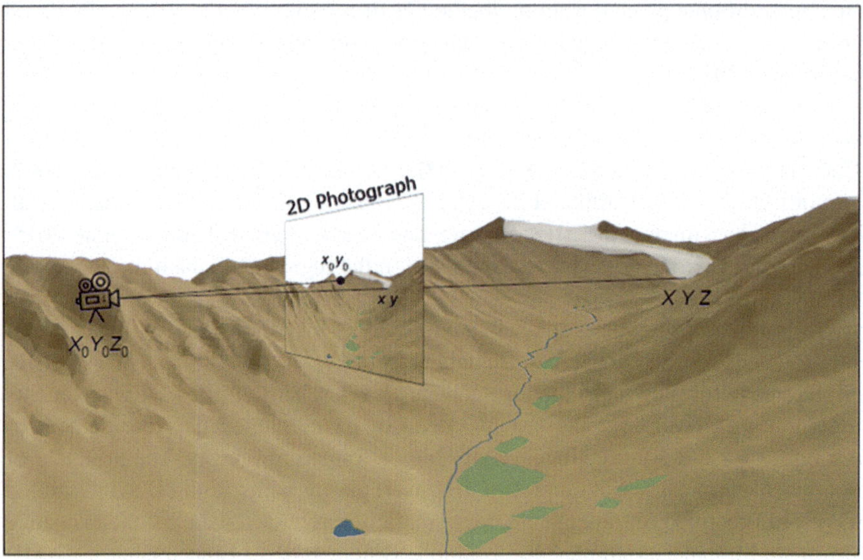

Fig. 2.9 Monoplotting and the collinearity equations. The camera (at 3D position $X_0Y_0Z_0$), a point on the 2D photograph (at xy), and the real world 3D point (at XYZ) all lie in a straight line. The collinearity equations solve the relationship between these points. x_0y_0 is the 2D coordinate of the line drawn from the projection centre of the camera through the centre of the photograph

the set of Eqs. 2.30 and 2.31, known as the collinearity equations:

$$x - x_0 = -c\frac{R_{11}(X - X_0) + R_{21}(Y - Y_0) + R_{31}(Z - Z_0)}{R_{13}(X - X_0) + R_{23}(Y - Y_0) + R_{33}(Z - Z_0)} \tag{2.30}$$

$$y - y_0 = -c\frac{R_{12}(X - X_0) + R_{22}(Y - Y_0) + R_{32}(Z - Z_0)}{R_{13}(X - X_0) + R_{23}(Y - Y_0) + R_{33}(Z - Z_0)} \tag{2.31}$$

where x and y are the coordinates of a point on the 2D photograph; x_0 and y_0 are the 2D coordinates of the line drawn from the projection centre of the camera through the centre of the photograph; X, Y, and Z are the coordinates of a 3D point in the real world; X_0, Y_0, and Z_0 are the coordinates of the camera at a 3D position in the real world; R_{11}–R_{33} are functions of the rotation angles of the camera about the X, Y, and Z axes; and c is the focal length of the camera (Stockdale 2017).

2.6.2 The WSL Monoplotting Tool

Over the last two decades, several software solutions have been developed on the basis of the monoplotting technique, e.g., OP-XFORM (Doytsher and Hall

1995), Georeferencing oblique terrestrial images (Corripio 2002), 3D Monoplotter (Mitishita et al. 2004), PRACTISE (Härer et al. 2016), RSG (Joanneum Research 2018), and Pic2Map (Produit 2019).

Most of above solutions prove very technical and non-intuitive. To meet the needs of end users in operational flexibility and user-friendliness, the MPT was developed at the Swiss Federal Institute for Forest, Snow and Landscape Research (WSL) (Bozzini et al. 2011). The tool is currently still under development.

The MPT contains a calibration routine, which computes the values of the collinearity equations and determines the extrinsic and intrinsic camera parameters. The routine functions with a set of four or more control points, whose relationship between the 2D oblique terrestrial image and the 3D real world is known (see Fig. 2.10). Once a calibration has been performed, the oblique image is orthorectified. The image can now be further analysed by importing or drawing polygonal vector shapes or lines along features of interest. In addition to supporting the export

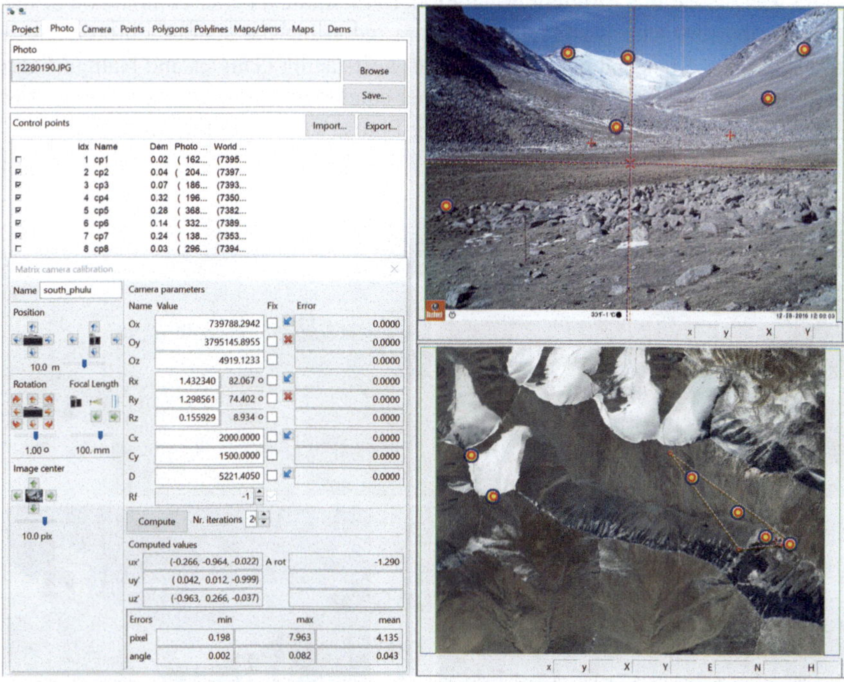

Fig. 2.10 The MPT's user interface. Control points (top left) are inserted on the 2D oblique terrestrial image (top right), with the help of an orthorectified image and underlying 3D DEM of the same area (bottom right). The "Matrix camera calibration" pane (bottom left) is the heart of the tool and solves the collinearity equations. Accurately placed control points show a good converging of the blue and red control circles. By eliminating the least accurate control points, the error values are reduced, until a suitable solution with the least error means is found

of such spatially referenced data to a GIS, already spatially referenced data can also be imported in the MPT, and superimposed on an image.

The accuracy of the MPT is affected by:

- The placement of control points (at least four control points are necessary)
- The resolution of a DEM
- The angle of incidence from the viewing location to the observed surface
- The lens distortion in the camera (Stockdale et al. 2015)

2.6.3 The Four Camera Stations

In order to monitor and validate on a local scale the snow cover extent, terrestrial cameras were set up at four locations in Leh Valley, namely Gyamtsa, Khakshal, Lungmar, and South Phulu (see Fig. 2.11 and Table 2.7). Care was taken that each camera was placed in a different orientation, to facilitate a better comparison of the snow distribution, with the Gyamtsa camera recording the south-facing Gyamtsa catchment area, Khakshal parts of the south-west-facing Gangles and Namgo Sago mountain area, Lungmar part of the Lungmar catchment and north-east-facing Stok range area and South Phulu the east-facing Phutse glacier and its river catchment

Fig. 2.11 Four camera stations: (**a**) Gyamtsa camera station and AWS; (**b**) Khakshal camera station and AWS; (**c**) Lungmar camera station; and (**d**) South Phulu camera station with the graduated snow stake with a length of 174 cm

Table 2.7 Details of the four locations and terrestrial cameras

	Gyamtsa	Khakshal	Lungmar	South Phulu
Area Extent SW:NE long., lat. (°)	77.53644, 34.21563: 77.59712, 34.27793	77.58975, 34.18346: 77.67349, 34.25729	77.32275, 33.88929: 77.61421, 34.19401	77.55207, 34.26881: 77.60458, 34.29152
Area (km^2)	18.74	23.13	354.68	5.79
Area Perimeter (km)	21.21	27.26	110.56	14.50
Area Maximum Elevation (m a.s.l.)	5730	5750	6090	5740
Area Minimum Elevation (m a.s.l.)	3980	3630	3140	4910
Area Mean Elevation (m a.s.l.)	4770	4700	4290	5300
Viewshed Area (km^2)	9.69	3.28	90.75	1.90
Camera	Raspberry Pi Camera Module v1	Raspberry Pi Camera Module v1	Bushnell Trophy Cam HD Essential	Bushnell Trophy Cam HD Essential
Resolution (MP)	5	5 (uncropped)	12	12
File format	JPEG	JPEG	JPEG	JPEG
Full-frame SLR Lens equivalent (mm)	35	35	28	28
Focal Ratio (F-Stop)	2.9	2.9	3.1	3.1
Camera long., lat. (°)	77.56869, 34.21563	77.58975, 34.18346	77.61421, 34.19401	77.60458, 34.26980
Camera Elevation (m a.s.l.)	4220	3660	4160	4900
Camera Height above ground (m)	0.08	1.66	0.11	2.71
Calibrated Camera Angle (°)	357.40	45.31	210.13	285.46
Calibrated Camera FOV (°)	55.54	31.24 (with cropped images)	43.28	42.58

area (BTW The South Phulu camera simultaneously took pictures of a 174 cm long graduated snow stake, see Fig. 2.11d, zoomed in portion of the snow stake).

The Gyamtsa and Lungmar cameras were used additionally for snow line analysis and validation during different seasons, and the more accessible lower elevation part of the South Phulu area was also used in detailed snow course research involving snow depth and SWE measurements (see Sect. 2.5.2).

Gyamtsa and Khakshal were part of the AWSs, with Gyamtsa additionally recording air temperature, barometric pressure and relative humidity data, and Khakshal also recording air temperature, barometric pressure, relative humidity, soil temperature, wind speed, wind direction, and the UV index.

Oblique terrestrial images were obtained from all the camera stations in an hourly interval in JPEG format and orthorectified using the MPT. In QGIS, the HMA 8 m filled DEM and location (viewpoint) of the stations, reprojected in UTM coordinates, were used to make camera viewsheds, that were clipped to the boundaries of the Field Of View (FOV) and photo extents of the cameras. The viewsheds were then further clipped to the extent of the watersheds of the locations, resulting in final viewsheds (see also Fig. 2.12). Since in the southern part of Leh Valley area the HMA DEM revealed a lot of missing pixels, TanDEM-X data was used to make the viewshed of the Lungmar location.

Subsequently, raster snow NDSI images of both MODIS and Landsat 8 OLI were clipped to the viewshed of each camera location and converted into polygon vector format. The polygons were then superimposed on orthorectified photographs of the same date in the MPT, in order to inspect and validate the snow cover congruence.

The Landsat polygons were further processed by converting them into polyline format, and imported in the MPT, to check and validate the snow lines against the orthorectified photos. For some snow line overviews the GNU Image Manipulation Program (GIMP) was also used. Table 2.8 provides more information about the achieved errors of the MPT related to the locations of the four camera stations. Since for the Lungmar location the TanDEM-X 12 m was used, the achieved error results show slightly higher compared to the other locations which used the higher resolution HMA 8-m DEM.

An overview of the sequence of the snow line delineation is shown in flowchart format in Fig. 2.13. Landsat snow lines, derived during different seasons, were superimposed and validated with photographs taken with the Gyamtsa and Lungmar camera stations.

The four cameras stations are described in detail in the following sections.

Gyamtsa Camera Station

The Gyamtsa (Tibetan: རྒྱ་རྩྭ , Wylie: *rgya rtswa*) camera station was part of the custom-made AWS, that operated from 1 December 2015 until 21 February 2017, at longitude 77°56869'E, latitude 34°21563'N at an elevation of 4220 m a.s.l., on a ridge in the Gyamtsa Valley, a western fork of the upper Leh Valley. The local meaning is a compound of the terms "gya"—large and "tsa"—fodder plants (Norman 2019), referring to large grazing areas containing fodder plants for the local cattle of Leh Valley. Figure 2.14a shows the AWS set-up with the camera and Fig. 2.14b provides a view of the south-facing Gyamtsa catchment area in the background.

The Gyamtsa site encompasses an area of 18.74 km^2, out of which the camera viewshed covers an area of 9.69 km^2, and extends from longitude 77°53644'E to 77°59712'E and latitude 34°21563'N to 34°27793'N. See Fig. 2.15 and Table 2.7 for more details.

The Gyamtsa custom-built outdoor AWS prototype was controlled by a low-power consuming RPi Model A+, a Linux-based single-board minicomputer, and

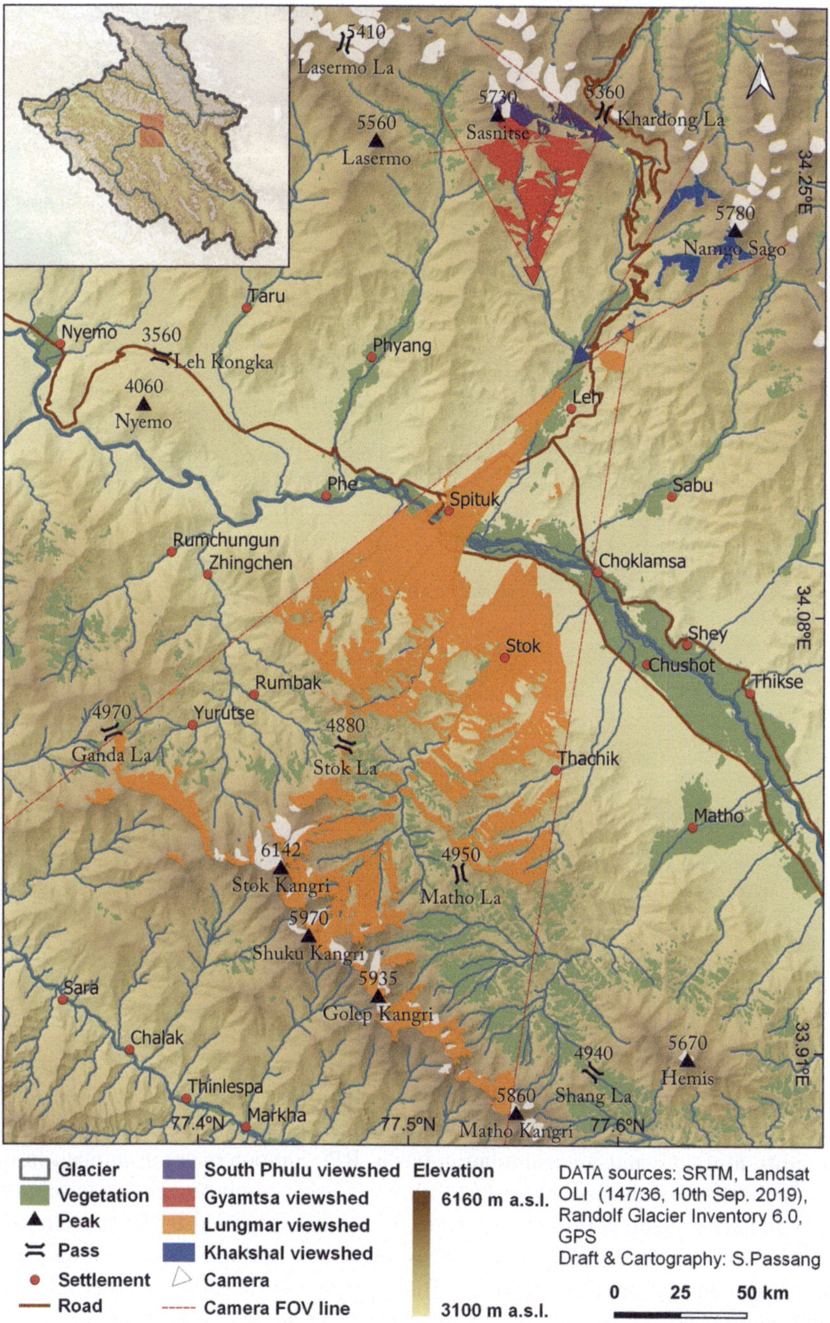

Fig. 2.12 Mounted terrestrial camera positions and viewsheds in Leh Valley

Table 2.8 Achieved theoretical 2D and 3D world errors of the MPT for the four locations of the camera stations

	Control points		2D error world (m)			3D error world (m)		
Location	Total	Final	Min	Max	Mean	Min	Max	Mean
Gyamtsa	8	6	1.2	10.5	6.2	1.2	10.7	6.6
Khakshal	8	5	0.1	7.0	2.9	0.1	7.5	3.2
Lungmar	12	6	0.6	20.9	8.7	0.6	24.9	9.5
South Phulu	8	6	0.1	11.1	3.5	0.1	11.8	3.8

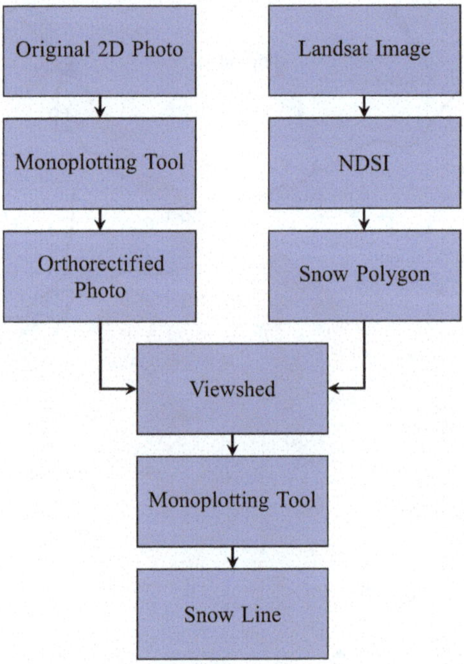

Fig. 2.13 Flowchart for delineating a snow line from a cloudless Landsat superimposed on a camera photo

integrated with an Adafruit BME280 environmental sensor (Adafruit Industries 2016) that recorded every hour the air temperature, barometric pressure, and relative humidity at the Gyamtsa accumulation basin. RPi computers are developed in the UK, with the mission to promote the teaching of basic computer science in schools. They quickly grew in popularity and today virtually every industry has developed a use for them, especially in robotics and weather monitoring. The station had been specifically chosen on a site devoid of Buddhist prayer flags or any other religious significance, and pre-researched for suitable photo angles of the basin. Table 2.9 illustrates the equipment used to build the station.

The station was also equipped with an RPi Camera Module that took 5 mega pixel High Definition (HD) photos towards the south-facing Gyamtsa catchment

(a) (b)

Fig. 2.14 Gyamtsa camera station (**a**) and AWS (**b**) with viewshed of the camera

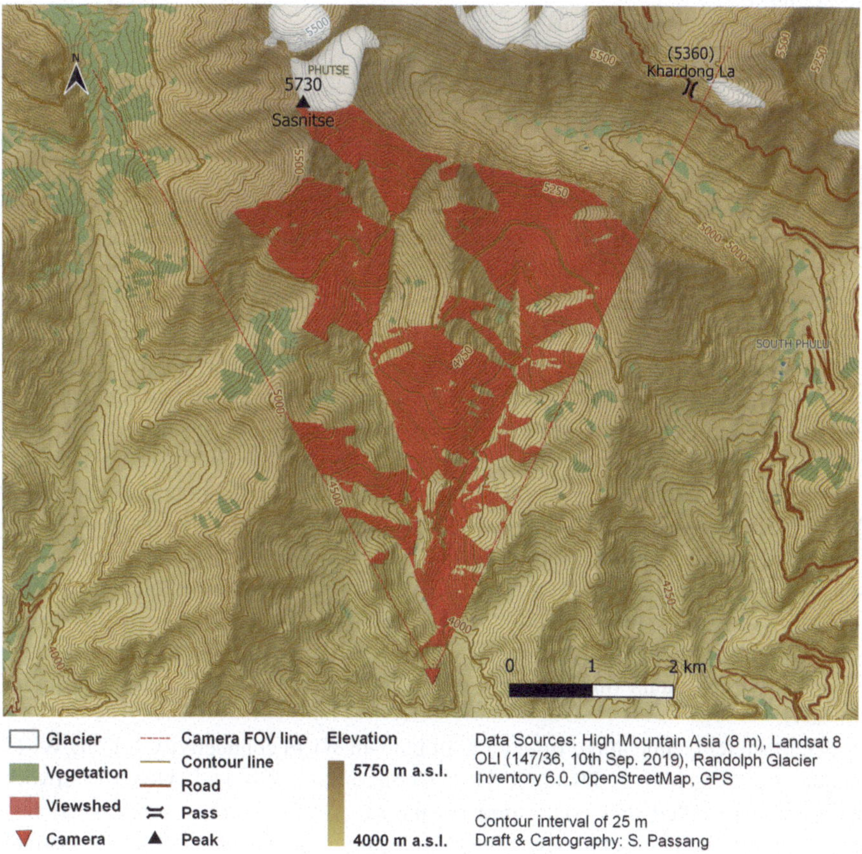

Fig. 2.15 Location and viewshed of Gyamtsa camera station

area, to monitor the snow line and capture daily cloud patterns. A Sleepy Pi (SPi), an Arduino based micro controller with a real time clock (RTC) (Spell Foundry

Table 2.9 Equipment used to build the Gyamtsa AWS

Item	Specification
Raspberry Pi 1 Model A+	Single-board computer
Micro SDHC card	OS and stores camera and sensor data
Raspberry Pi Camera Module v1	5 mega pixel HD photos
Adafruit BME280	Temperature, barometric pressure and humidity sensor
Adafruit 1/2 breadboard	BME280 fixing and wiring
Sleepy Pi	Arduino microcontroller for RTC and time loop for recordings
CR1632 lithium coin battery	Power supply of RTC
Wireless USB adapter	Connects wirelessly to the RPi
Solar Panel (2x)	12 V 10 W
Solar charge controller and rechargeable batteries	12 V/7 Ah
USB car power adapter	Converts 12 V to 5 V with 1 A current

2016), provided an hourly time interval to record the photos and sensor values and simultaneously saved system power. The RTC was powered by a Lithium coin battery and provided accurate date and time details for the offline location.

Initially one, later two parallel connected 12 V 10 W solar panels supplied power to the RPi, using a rechargeable 12 V/7 Ah battery and solar charge controller. The power was converted from 12 V to 5 V (the operating voltage of the electronic components) with a USB car power adapter. An insulated wooden box painted black was used to secure battery and solar charge controller from extreme weather, and a wooden box painted white with holes to allow airflow functioned as a mini Stevenson screen housing the RPi, RPi Camera Module, SPi and BME280 sensor.

A customised operating system, Sleepy Pi OS, based on Raspbian, was installed on a 32 GB Micro Secure Digital High Capacity (SDHC) card, and dedicated code developed for the configuration of the RPi, RPi Camera Module, SPi, sensor and sensor data recording time loop. The card also stored the camera and sensor data and allowed during data collection in the field, a convenient swap with another identically programmed 32 GB SDHC card. Programming was done in a Secure Shell (SSH) command line terminal, using the program PuTTY, a client program for SSH.

Since the RPi came at the time without built-in Wi-Fi connectivity, a USB Wi-Fi adapter was attached. A Dynamic Host Configuration Protocol (DHCP) network server was installed and configured on the OS, to facilitate a wireless ad hoc connection to the RPi in the field. The mobile app Fing was used to find the correct DHCP lease Internet Protocol (IP) address issued by the DHCP server, wirelessly connect to the RPi for the initial set-up of the correct angle of the camera, for occasional data checks, enable code modifications in the field, and the starting or shutting down of the whole system.

(a) (b)

Fig. 2.16 Khakshal camera station (**a**) and AWS (**b**) with viewshed of the camera

Khakshal Camera Station

The Khakshal (Tibetan: ཁག་ཤལ་ , Wylie: *khak shal*) camera station is part of the custom-made AWS, that has been operating since 28 May 2016 till the present, at longitude 77°58975'E, latitude 34°18346'N at an elevation of 3660 m a.s.l. in Khakshal, a village in Leh Valley about 2.5 km north of Leh town. The local meaning is a compound of the terms "khak"—part and "shal"—rinse (Norman 2019), referring to a part of the area that got washed away during a flood in the past. Figure 2.16a shows the camera front view and Fig. 2.16b shows the AWS set-up and a view of the Khakshal watershed area with the south-west-facing Gangles and Namgo Sago peak area in the centre background.

The Khakshal site encompasses an area of 23.13 km^2, out of which the camera viewshed covers an area of 3.28 km^2, and extends from longitude 77°58975'E to 77°67349'E and latitude 34°18346'N to 34°25729'N. See Fig. 2.17 and Table 2.7.

The Khakshal custom-built outdoor AWS (see Fig. 2.18) is one of the first electronic full-fledged stations in the region and is equipped with a SparkFun Weather Shield containing air temperature, relative humidity, and barometric environmental sensors (SparkFun Electronics 2016a,b), Particle Photon Wi-Fi microcontroller (Particle 2016) and an RPi 3 Model B Linux-based single-board minicomputer, all placed within a self-constructed Stevenson screen. An RPi Camera Module and UV index sensor were also integrated in the station. An anemometer and wind vane were mounted on a pole five and half meter above the ground. Moreover, a soil temperature sensor was installed at 10 cm underground. For a list of the equipment see Table 2.10. Figure F shows an assemblage of photos of the AWS build procedure.

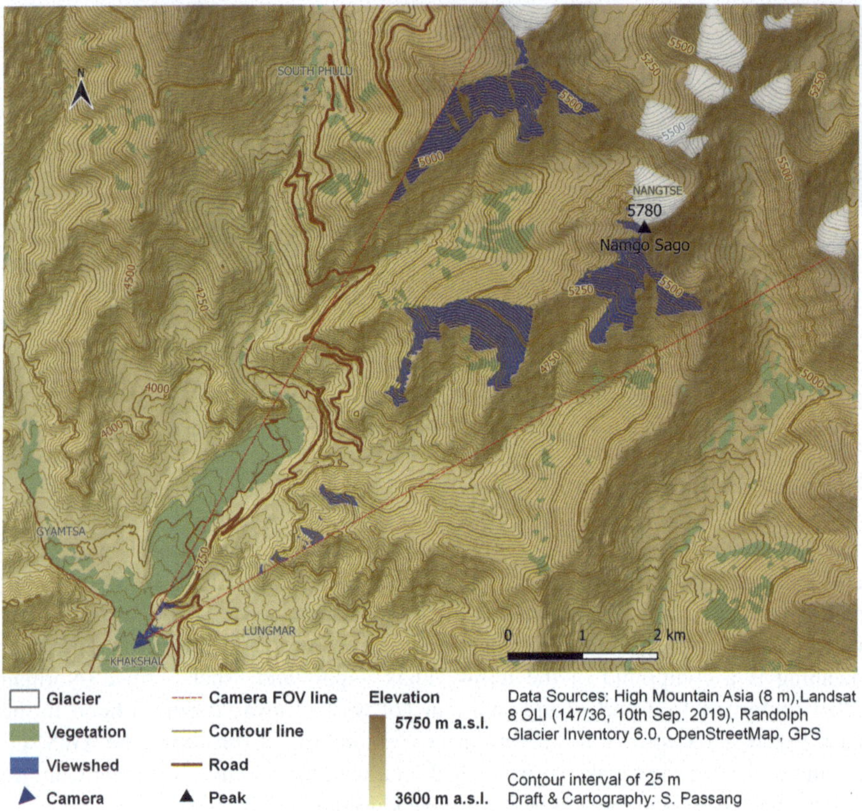

Fig. 2.17 Location and viewshed of Khakshal camera station

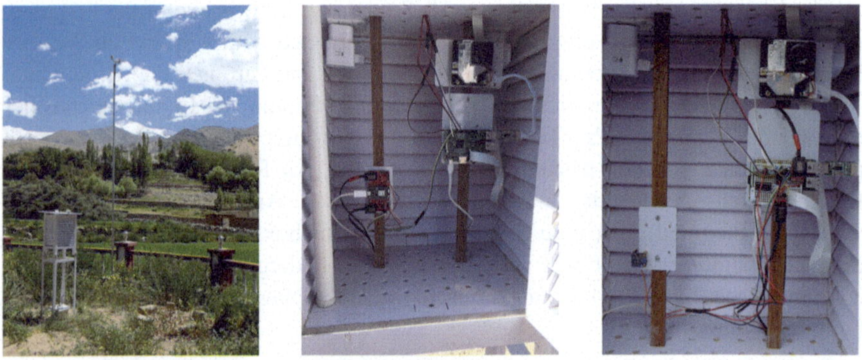

(a) (b) (c)

Fig. 2.18 Khakshal AWS. (**a**) Khakshal AWS set-up location. (**b**) Electronic components and sensors. (**c**) Upgraded components and sensors

Table 2.10 Equipment used to build the Khakshal AWS

Item	Specification
Raspberry Pi 3 Model B	Single-board computer
Micro SDHC card	OS and stores camera and sensor data
Raspberry Pi Camera Module v2	8 mega pixel HD photos
Particle Photon kit	Wi-Fi IoT (Internet of Things) microcontroller
SparkFun Weather Shield	Add-on PCB (Printed Circuit Board) for the Particle Photon Kit, featuring temperature-, barometric pressure- & humidity sensors
Soil temperature sensor DS18B20	Soil temperature
Anemometer	Wind speed sensor
Wind vane	Wind direction sensor
UV sensor SI1145	UV Index

Experience from the Gyamtsa AWS prototype was incorporated in the Khakshal AWS, and after many trials and errors and the constant updating of the software, specific electronic components were chosen for it.

In order to avoid wrong readings due to measuring instruments being directly exposed to the elements, in the nineteenth century Scottish civil engineer Thomas Stevenson developed a method to make an outside weather station with the instruments housed inside in a small "shelter", now known as a Stevenson screen after his name (Stevenson 1866).

A Stevenson screen is built using double sided slanted louvres. If wind hits it cannot blow through straight on the instruments, it has to make a double turn and while doing so it loses its energy so the climate inside is without wind chill. The screen is usually made of wood and painted white to reflect the heat from sunlight. The top and bottom of the screen are double with drilled holes for air circulation, and the outer roof is slanted, so any snow or rainfall can fall or drain off on the backside. The screen was constructed at such a distance from the ground that the thermometer sensor could be placed at a required minimum height of 1.25 m from the ground.

The camera was mounted in a small wooden box on the roof of the station and makes a picture every hour in a north-eastern direction, towards the south-west facing Gangles and Namgo Sago peak area. Namgo Sago is one of the peaks of the Ladakh range, with an elevation of 5750 m a.s.l. This to monitor the snow line and to capture daily cloud patterns. The camera timing interval was adapted to each season in accordance with available day light, for example the timing for peak summer was set-up from 5 a.m. to 8 p.m., and for peak winter from 7 a.m. to 5 p.m.

The UV sensor was mounted on top of the station in a small wooden box similar to the camera box. The box has a hinge, so the angle of the box can be controlled seasonally. In this way sunrays hit the sensor more optimally throughout the year, according to the position of sun.

On the SparkFun Weather Shield several components were soldered to facilitate the wiring of some sensors. Incoming sensor data of the Weather Shield can be

Table 2.11 Weather sensors of the SparkFun Weather Shield

Sensor	Specification
HTU21D humidity and temperature sensor	Temperature accuracy: ±0.3 °C, Humidity accuracy: ±2%
MPL3115A2 barometric pressure sensor	Pressure accuracy: ±0.05kPa, Altitude accuracy: ±0.3m
Temperature sensor—waterproof (DS18B20)	Temperature accuracy: ±0.5 °C

downloaded from the serial port of the Photon to the RPi, using a Python script. The Wi-Fi Internet of Things (IoT) possibilities of the Photon have a lot of advantages while on a stable internet connection, however, not always available in Ladakh. Table 2.11 lists more information about the sensors of the SparkFun Weather Shield.

Initially an air temperature difference of about 3.8°C was observed, due to heat emitted by the Photon. Therefore a calibration was performed using a generic industrial grade thermometer. Relative humidity, barometric pressure and soil temperature values were showing normal and needed no calibration.

An extension underneath the AWS was manufactured, a small wooden box equipped with an RPi 3 Model B and RPi Camera Module. During the snow season it records the snowboard with inserted ruler, to measure the snow depth. For an example of a snowboard picture taken by the RPi Camera Module, see Fig. 2.6b.

In 2019 an upgrade was performed to the set-up of the AWS. Some changes were necessary due to ongoing problems with the internet connectivity in Ladakh, causing infrequent readings/failures of the Photon and SparkFun Weather Shield. The Photon and Weather Shield were removed and replaced by a customised self-soldered weather board, mounted directly on top of the RPi. The RPi now directly collects all the incoming sensor data, without the necessity of the internet. To measure particulate matter (PM) pollution with a diameter of 2.5 and 10 μm or less, a Nova PM sensor was installed as well, directly connected to the RPi via a USB connector (Nova Fitness Co., Ltd. 2019; Raspberry Pi Foundation 2019). See Fig. 2.18b, prior to the upgrade of the AWS, and Fig. 2.18c for the upgraded weather components.

Concerning the software, Raspberry Pi OS (previously called Raspbian), a computer operating system for the RPi, was installed on a 32 GB Micro SDHC card, and appropriate code developed for the configuration of the RPi, RPi Camera Module, Photon, SparkFun Weather Shield, Wi-Fi, sensors and the sensor data recording time loop. Most of the programming was again done in a SSH command line terminal with PuTTY. The Photon and Weather Shield were programmed in Wiring, a computer programming language written in C/C++.

Prior to the upgrade, weather sensor data were recorded every 5 minutes and stored in a MySQL database. The current set-up stores every 5 minutes in a MariaDB database, which is a community-developed, open-source format and multi-configurable with scripts and algorithms, providing more flexibility for the AWS project in the long term.

(a) (b)

Fig. 2.19 Lungmar camera station (**a**) and viewshed of the camera (**b**)

Electric power for the AWS is provided from the mains electricity grid. For backup power there is a 12 V/150 Ah battery inside the nearby house that charges from the mains and three solar panels. The station is therefore also indirectly solar-powered.

Lungmar Camera Station

The Lungmar (Tibetan: ལུང་དམར་ , Wylie: *lung dmar*) camera station operated from 24 November 2015 until 30 May 2016, at longitude 77°61421'E, latitude 34°19401'N at an elevation of 4160 m a.s.l. in Lungmar Valley, an eastern fork of upper Leh Valley to the east side of Gonpa village. The local meaning is a contraction of the terms "lungpa"—valley and "marpo"—red (Norman 2019), referring to large parts of the valley having red rock formations. Figure 2.19a shows the camera front view and Fig. 2.19b provides a view of the vast Lungmar and Stok range watershed areas in the background.

The Lungmar site encompasses an area of 354.68 km^2, out of which the camera viewshed covers an area of 90.75 km^2, and extends from longitude 77°32275'E to 77°61421'E and latitude 33°19401'N to 34°88929'N. For more details see Fig. 2.20 and Table 2.7.

A Bushnell Trophy Camera is a wildlife camera mainly used for taking wildlife pictures, hunting purposes, etc. However due to its high-quality output and reliable performance under the most extreme weather conditions, it was very suitable for the Lungmar site. The camera is equipped with a timestamp. Energy lithium AA batteries were specifically used because of their long-lasting performance in extreme weather conditions. The camera was housed in an insulated wooden box and installed on a mountain slope, and a stone structure was built around it to protect it from the weather elements.

The camera recorded 12 megapixel photographs of the changing patterns of the snow line on the north-facing slopes of the Stok range to the south of the Indus river.

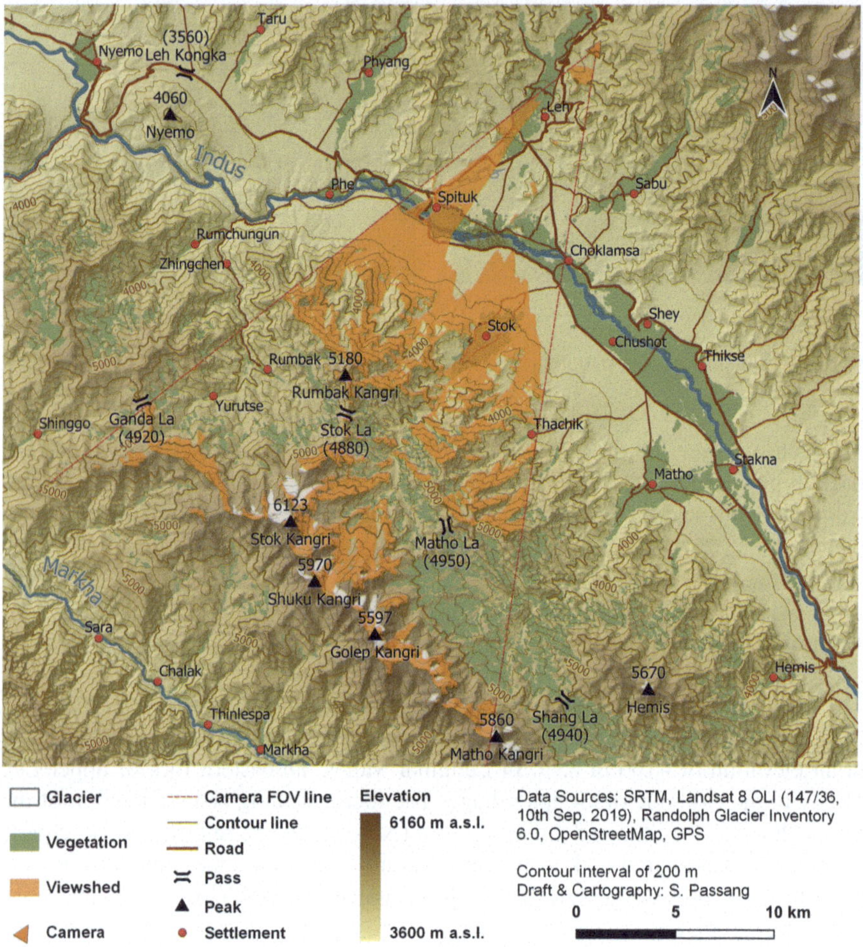

Fig. 2.20 Location and viewshed of Lungmar camera station

The data available from the Lungmar camera station are of limited duration due to the relocation of the camera to the South Phulu site.

South Phulu Camera Station

The South Phulu (Tibetan: ཕུ་ལུ་, Wylie: *phu lu*) camera station operated from 15 December 2016 until 23 May 2017, at longitude 77°60458'E, latitude 34°26980'N at an elevation of 4900 m a.s.l., on the south-facing slope of upper Leh Valley near the Khardong La pass. The local meaning refers to the term "phulu"—shepherd's hut (Norman 2019), pointing to the abandoned shepherds' huts at the site.

(a) (b)

Fig. 2.21 South Phulu camera station (**a**) and viewshed of the camera (**b**)

Figure 2.21a shows the camera front view and Fig. 2.21b shows the camera placed in a stone-studded enclosure on a glacial boulder, with the South Phulu watershed area and Phutse glacier in the background.

The South Phulu site encompasses an area of 5.79 km^2, out of which the camera viewshed covers an area of 1.90 km^2, and extends from longitude 77°55207'E to 77°60458'E and latitude 34°26881'N to 34°29152'N. See Fig. 2.22 and Table 2.7.

The Bushnell Trophy Camera of Lungmar was relocated to the South Phulu location. The camera was housed in an insulated wooden box and installed on a glacial boulder, and a stone structure was built around it to protect it from the weather elements.

The camera recorded 12 megapixel photographs of a graduated snow stake, to investigate the spatio-temporal patterns of the snow cover, snow depth and snowfall interception in the vicinity of the camera, and validating the patterns of the snow cover and Phutse glacier in the distant range of the South Phulu watershed area. The stake had a height of 174 cm and was painted red and green at a 5 cm interval. It provided an alternative method to the traditional survey method using snow plots and prevented data gaps for this period. The image software Digimizer was used to calibrate the stake height in the photographs with a precision of 1 cm.

Time-lapse photography in combination with snow stakes is a proven technique to study snow cover, used all over the world (Parajka et al. 2012; Garvelmann et al. 2013; Hedrick and Marshall 2014; Dong and Menzel 2017; Bongio et al. 2021).

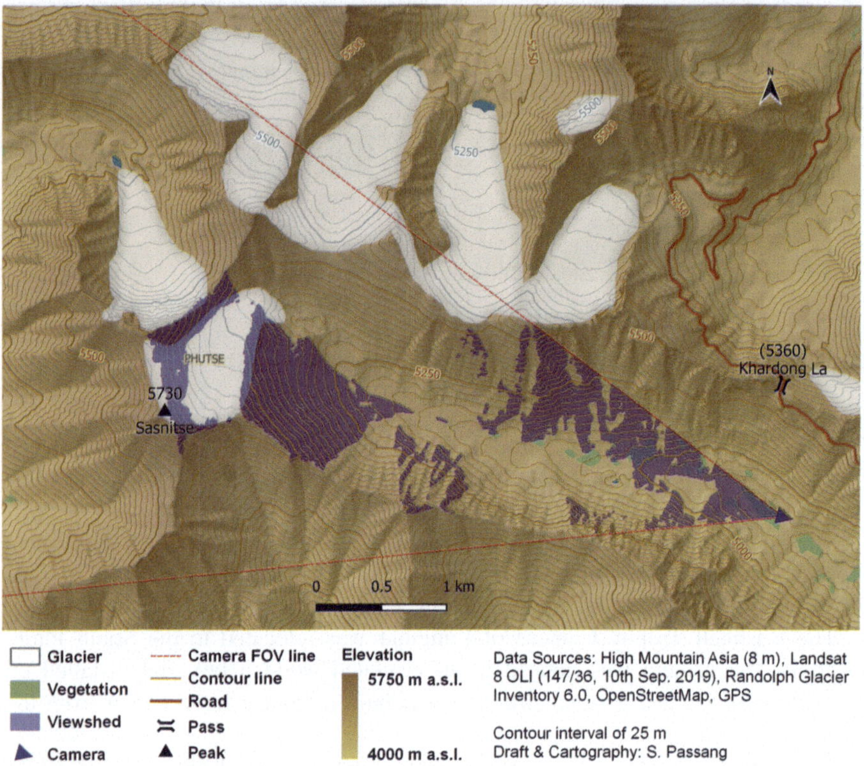

Fig. 2.22 Location and viewshed of South Phulu camera station

References

Abdul-Niby M, Farhat M, Abdullah M, Nazzal A (2017) A Low cost automated weather station for real time local measurements. Eng Technol Appl Sci Res 7(3):1615–1618. http://www.etasr.com/index.php/ETASR/article/view/1187

Adafruit Industries (2016) Open-source hardware company. https://www.adafruit.com, accessed 1 July 2016

Akaike H (1973) Information theory and an extension of the maximum likelihood principle. In: Proceeding of the second international symposium on information theory, Akadémiai Kiadó, Budapest, Hungary, pp 267–281

Akaike H (1974) A new look at the statistical model identification. IEEE Trans Autom Control 19(6):716–723. https://doi.org/10.1109/TAC.1974.1100705

Akpanta A, Okorie I, Okoye N (2015) SARIMA modelling of the frequency of monthly rainfall in Umuahia, Abia State of Nigeria. Am J Math Stat 5(2):82–87. https://doi.org/10.5923/j.ajms.20150502.05

Bernard É, Friedt JM, Tolle F, Griselin M, Martin G, Laffly D, Marlin C (2013) Monitoring seasonal snow dynamics using ground based high resolution photography (Austre Lovénbreen, Svalbard, 79°N). ISPRS J Photogramm Remote Sens 75:92–100. https://doi.org/10.1016/j.isprsjprs.2012.11.001

Bongio M, Arslan AN, Tanis CM, De Michele C (2021) Snow depth time series retrieval by time-lapse photography: Finnish and Italian case studies. Cryosphere 15(1):369–387. https://doi.org/10.5194/tc-15-369-2021

Box GE, Pierce DA (1970) Distribution of residual autocorrelations in autoregressive-integrated moving average time series models. J Am Stat Assoc 65(332):1509–1526. https://doi.org/10.1080/01621459.1970.10481180

Box GE, Jenkins GM, Reinsel GC (2008) Time series analysis: forecasting and control. Wiley, Hoboken. https://doi.org/10.1002/9781118619193

Bozzini C, Conedera M, Krebs P (2011) A new tool for obtaining cartographic georeferenced data from single oblique photos. In: Proceedings of the XXIIIrd international CIPA Symposium, Prague, Czech Republic, vol 1216. WSL Swiss Federal Research Institute, Bellinzona, p 6. https://www.wsl.ch/fileadmin/user_upload/WSL/Projekte/Evoluzione_paesaggio/Bozzini_CIPA2011_en.pdf

Brockwell PJ, Davis RA (1991) Time series: theory and methods. Springer, New York. https://doi.org/10.1007/978-1-4419-0320-4

Brombierstäudl D, Schmidt S, Nüsser M (2021) Distribution and relevance of aufeis (icing) in the Upper Indus Basin. Sci Total Environ 780:146604. https://doi.org/10.1016/j.scitotenv.2021.146604

Burnham KP, Anderson DR (1998) Practical use of the information-theoretic approach. In: Model selection and inference: a practical information-theoretic approach. Springer, New York, pp 75–117. https://doi.org/10.1007/978-1-4757-2917-7_3

Bushnell Corporation (2016) Sporting optics and outdoor products. https://www.bushnell.com, accessed 1 July 2016

Cavanaugh JE (1997) Unifying the derivations for the Akaike and corrected Akaike information criteria. Stat Probab Lett 33(2):201–208. https://doi.org/10.1016/S0167-7152(96)00128-9

Chatfield C, Xing H (2019) The analysis of time series: an introduction with R. CRC Press, Boca Raton. https://doi.org/10.1201/9781351259446

Chawsheen TA, Broom M (2017) Seasonal time-series modeling and forecasting of monthly mean temperature for decision making in the Kurdistan Region of Iraq. J Stat Theory Pract 11(4):604–633. https://doi.org/10.1080/15598608.2017.1292484

Chokmani K, Dever K, Bernier M, Gauthier Y, Paquet LM (2010) Adaptation of the SNOWMAP algorithm for snow mapping over eastern Canada using Landsat-TM imagery. Hydrol Sci J–Journal des Sciences Hydrologiques 55(4):649–660. https://doi.org/10.1080/02626661003747374

Christiansen HH (2001) Snow-cover depth, distribution and duration data from northeast Greenland obtained by continuous automatic digital photography. Ann Glaciol 32:102–108. https://doi.org/10.3189/172756401781819355

Coban V, Guler E, Kilic T, Kandemir SY (2021) Precipitation forecasting in Marmara region of Turkey. Arab J Geosci 14(2):10. https://doi.org/10.1007/s12517-020-06363-x

Cohen J (1960) A coefficient of agreement for nominal scales. Educ Psychol Meas 20(1):37–46. https://doi.org/10.1177/001316446002000104

Corripio JG (2002) Georeferencing oblique terrestrial photography. Zürich, Switzerland. https://www.arolla.ethz.ch/IDL/georefusermanual.pdf

Cryer JD, Chan KS (2008) Time series analysis: with applications in R, vol 2. Springer, New York. https://doi.org/10.1007/978-0-387-75959-3

Dickey DA, Fuller WA (1979) Distribution of the estimators for autoregressive time series with a unit root. J Am Stat Assoc 74(366a):427–431. https://doi.org/10.2307/2286348

Dimri T, Ahmad S, Sharif M (2020) Time series analysis of climate variables using seasonal ARIMA approach. J Earth Syst Sci 129(1):16. https://doi.org/10.1007/s12040-020-01408-x

Dong C (2016) Assessing the availability of remote sensing, hydrological modeling and in situ observations in snow cover research. PhD thesis, Heidelberg University, Heidelberg. https://doi.org/10.11588/heidok.00021603

Dong C, Menzel L (2016) Producing cloud-free MODIS snow cover products with conditional probability interpolation and meteorological data. Remote Sens Environ 186:439–451. https://doi.org/10.1016/j.rse.2016.09.019
Dong C, Menzel L (2017) Snow process monitoring in montane forests with time-lapse photography. Hydrol Process 31(16):2872–2886. https://doi.org/10.1002/hyp.11229
Doytsher Y, Hall JK (1995) FORTRAN programs for coordinate resection using an oblique photograph and high-resolution DTM. Comput Geosci 21(7):895–905. https://doi.org/10.1016/0098-3004(95)00023-2
Eder K, Aumann G (1996) An integrated system of digital monoplotting and DTM modelling for forestry applications. Int Arch Photogramm Remote Sens 31:221–225. https://www.isprs.org/proceedings/XXXI/congress/part7/221_XXXI-part7.pdf
Everitt B, Skrondal A (2010) The Cambridge Dictionary of Statistics, vol 4. Cambridge University, Cambridge
Farinotti D, Magnusson J, Huss M, Bauder A (2010) Snow accumulation distribution inferred from time-lapse photography and simple modelling. Hydrol Process 24(15):2087–2097. https://doi.org/10.1002/hyp.7629
Floyd W, Weiler M (2008) Measuring snow accumulation and ablation dynamics during rain-on-snow events: innovative measurement techniques. Hydrol Process 22(24):4805–4812. https://doi.org/10.1002/hyp.7142
Fuller WA (1995) Introduction to statistical time series. Wiley, Hoboken. https://doi.org/10.1002/9780470316917
Gabellieri N, Watkins C (2019) Measuring long-term landscape change using historical photographs and the WSL Monoplotting Tool. Landsc Hist 40(1):93–109. https://doi.org/10.1080/01433768.2019.1600946
Gaffey C, Bhardwaj A (2020) Applications of unmanned aerial vehicles in cryosphere: latest advances and prospects. Remote Sens 12(6):948. https://doi.org/10.3390/rs12060948
Gafurov A, Bárdossy A (2009) Cloud removal methodology from MODIS snow cover product. Hydrol Earth Syst Sci 13(7):1361–1373. https://doi.org/10.5194/hess-13-1361-2009
Gafurov A, Lüdtke S, Unger-Shayesteh K, Vorogushyn S, Schöne T, Schmidt S, Kalashnikova O, Merz B (2016) MODSNOW-Tool: an operational tool for daily snow cover monitoring using MODIS data. Environ Earth Sci 75(14):1078. https://doi.org/10.1007/s12665-016-5869-x
Gao Y, Xie H, Yao T, Xue C (2010) Integrated assessment on multi-temporal and multi-sensor combinations for reducing cloud obscuration of MODIS snow cover products of the Pacific Northwest USA. Remote Sens Environ 114(8):1662–1675. https://doi.org/10.1016/j.rse.2010.02.017
Garvelmann J, Pohl S, Weiler M (2013) From observation to the quantification of snow processes with a time-lapse camera network. Hydrol Earth Syst Sci 17(4):1415. https://doi.org/10.5194/hess-17-1415-2013
German Aerospace Center (DLR) (2023) TerraSAR-X add-on for Digital Elevation Measurement (TanDEM-X). https://www.dlr.de/hr/en/desktopdefault.aspx/tabid-2317, accessed 15 July 2023
Gray DM, Male DH (1981) Handbook of Snow: Principles, Processes, Management & Use. Pergamon Press, Oxford, UK
Gurung DR, Giriraj A, Aung KS, Shrestha BR, Kulkarni AV (2011) Snow-cover mapping and monitoring in the Hindu Kush-Himalayas. Tech. rep., International Centre for Integrated Mountain Development (ICIMOD), Kathmandu. https://doi.org/10.53055/ICIMOD.550
Hall DK, Riggs GA, Salomonson VV (1995) Development of methods for mapping global snow cover using moderate resolution imaging spectroradiometer data. Remote Sens Environ 54(2):127–140. https://doi.org/10.1016/0034-4257(95)00137-P
Hall DK, Riggs GA, Salomonson VV, Barton J, Casey K, Chien J, DiGirolamo N, Klein A, Powell H, Tait A (2001) Algorithm theoretical basis document (ATBD) for the MODIS snow and sea ice-mapping algorithms. NASA/Goddard Space Flight Center 45:14–28. https://eospso.gsfc.nasa.gov/sites/default/files/atbd/atbd_mod10.pdf

Hall DK, Riggs GA, Salomonson VV, DiGirolamo NE, Bayr KJ (2002) MODIS snow-cover products. Remote Sens Environ 83(1–2):181–194. https://doi.org/10.1016/S0034-4257(02)00095-0

Hamed KH, Rao AR (1998) A modified Mann-Kendall trend test for autocorrelated data. J Hydrol 204(1–4):182–196. https://doi.org/10.1016/S0022-1694(97)00125-X

Haq MA, Baral P, Yaragal S, Rahaman G (2020) Assessment of trends of land surface vegetation distribution, snow cover and temperature over entire Himachal Pradesh using MODIS datasets. Nat Resour Model 33(2):e12262. https://doi.org/10.1111/nrm.12262

Härer S, Bernhardt M, Schulz K (2016) PRACTISE–Photo rectification and classification software (V. 2.1). Geosci Model Dev 9(1):307–321. https://doi.org/10.5194/gmd-9-307-2016

Harvey AC, Shephard N (1993) 10 Structural time series models. In: Econometrics, handbook of statistics, vol 11. Elsevier, Amsterdam, pp 261–302. https://doi.org/10.1016/S0169-7161(05)80045-8

Hauksson BÞ (2017) Estimation of regional snow-line elevations in Norway. In: Testing a method for estimating RSLE and melt slope analysis in Norway. Master's thesis, University of Oslo, Oslo. http://urn.nb.no/URN:NBN:no-60729

Hedrick AR, Marshall HP (2014) Automated snow depth measurements in avalanche terrain using time-lapse photography. In: 2014 International Snow Science Workshop, International Snow Science Workshop, Banff, Canada. https://doi.org/10.13140/2.1.2049.6646

van Herwijnen A, Berthod N, Simenhois R, Mitterer C (2013) Using time-lapse photography in avalanche research. In: 2014 International Snow Science Workshop, International Snow Science Workshop, Grenoble, Switzerland, pp 950–954

Huintjes E, Sauter T, Schröter B, Maussion F, Yang W, Kropáček J, Buchroithner M, Scherer D, Kang S, Schneider C (2015) Evaluation of a coupled snow and energy balance model for Zhadang glacier, Tibetan Plateau, using glaciological measurements and time-lapse photography. Arct Antarct Alp Res 47(3):573–590. https://doi.org/10.1657/AAAR0014-073

Hyndman RJ (2015) Discussion of "High-dimensional autocovariance matrices and optimal linear prediction". Electron J Stat 9(1):792–796. https://doi.org/10.1214/14-EJS953

Hyndman RJ, Athanasopoulos G (2021) Forecasting: principles and practice. OTexts, Melbourne, Australia. https://otexts.com/fpp3

Ihlen V (2019) Landsat 8 (L8): Data Users Handbook. US Geological Survey, John W. Powell National Center, Restion. https://www.usgs.gov/landsat-missions/landsat-8-data-users-handbook

Jain SK, Goswami A, Saraf AK (2009) Role of elevation and aspect in snow distribution in Western Himalaya. Water Resour Manag 23(1):71–83. https://doi.org/10.1007/s11269-008-9265-5

Jet Propulsion Laboratory NASA (2023) Shuttle Radar Topography Mission (SRTM). https://www2.jpl.nasa.gov/srtm/, accessed 15 July 2023

Joanneum Research (2018) Remote Sensing Software Graz. https://www.joanneum.at/fileadmin/user_upload/imported/DIGITAL/Downloads/Remote_Sensing_Software_Graz.pdf, accessed 1 July 2016

Karamouz M, Zahraie B (2004) Seasonal streamflow forecasting using snow budget and El Niño-southern oscillation climate signals: Application to the Salt River Basin in Arizona. J. Hydrol. Eng. 9(6):523–533. https://doi.org/10.1061/(ASCE)1084-0699(2004)9:6(523)

Karimi S, Niksokhan MH, Karimi S (2016) Modeling snow cover area and predicting its changes in Haraz catchment. Imaging 2(4):450–455

Kendall MG (1948) Rank correlation methods. Charles Griffin and Co. Ltd., London, UK

Krajčí P, Holko L, Perdigão RA, Parajka J (2014) Estimation of regional snowline elevation (RSLE) from MODIS images for seasonally snow covered mountain basins. J Hydrol 519:1769–1778. https://doi.org/10.1016/j.jhydrol.2014.08.064

Kumar M, Tiwari RK, Kumar K, Rautela KS (2022) Prediction of Seasonal Snow Accumulation and Depletion by SARIMA Model using MODIS data. In: AQUA—Water Infrastructure, Ecosystems and Society, p 15. https://doi.org/10.21203/rs.3.rs-2024923/v1

Kwiatkowski D, Phillips PC, Schmidt P, Shin Y (1992) Testing the null hypothesis of stationarity against the alternative of a unit root: how sure are we that economic time series have a unit root? J Econ 54(1–3):159–178. https://doi.org/10.1016/0304-4076(92)90104-Y

Liu J, Chen R, Wang G (2015) Snowline and snow cover monitoring at high spatial resolution in a mountainous river basin based on a time-lapse camera at a daily scale. J Mt Sci 12(1):60–69. https://doi.org/10.1007/s11629-013-2842-y

Ljung GM, Box GE (1978) On a measure of lack of fit in time series models. Biometrika 65(2):297–303. https://doi.org/10.1093/biomet/65.2.297

Lorentzen T (2014) Statistical analysis of temperature data sampled at Station-M in the Norwegian Sea. J Mar Syst 130:31–45. https://doi.org/10.1016/j.jmarsys.2013.09.009

Makarovic B (1973) Digital mono-plotters. ITC J 4:583–599

Mann HB (1945) Nonparametric tests against trend. Econometrica: J. Econometric Soc 13(3):245–259. https://doi.org/10.2307/1907187

Marco C, Claudio B, Ueli R, Thalia B, Patrik K (2018) Using the Monoplotting Technique for Documenting and Analyzing Natural Hazard Events. In: Natural hazards-risk assessment and vulnerability reduction. IntechOpen Limited, Berlin/Heidelberg, chap 7, pp 107–123. https://doi.org/10.5772/intechopen.77321

Martinec J, Sevruk B (1992) Snow cover measurements and areal assessment of precipitation and soil moisture. Operational Hydrology Report, 35 WMO-No. 749:312. https://library.wmo.int/idurl/4/33781

McMurry TL, Politis DN (2010) Banded and tapered estimates for autocovariance matrices and the linear process bootstrap. J Time Ser Anal 31(6):471–482. https://doi.org/10.1111/j.1467-9892.2010.00679.x

Mitishita E, Machado A, Habib A, Gonçalves G (2004) 3D monocular restitution applied to small format digital airphoto and laser scanner data. In: Proceedings of Commission III, ISPRS Congress, Istanbul, International Society for Photogrammetry and Remote Sensing, Hannover, Germany. https://www.isprs.org/proceedings/xxxv/congress/comm3/papers/243.pdf

MODIS Reprojection Tool (2011) MODIS Reprojection Tool User's Manual (Release 4.1). Land Processes DAAC USGS Earth Resources Observation and Science (EROS) Center, Sioux Falls, USA

National Snow & Ice Data Center, CIRES (2023) High Mountain Asia (HMA). https://nsidc.org/data/highmountainasia, accessed 15 July 2023

Norman R (2019) Rang-Skat: A Ladakhi-English Dictionary. Edited by Henk Thoma. Melong Publications, Leh, Ladakh India

Nova Fitness Co, Ltd (2019) High-tech enterprise. http://inovafitness.com/en/a/index.html, accessed 1 March 2019

Panda S, Rao M, Thenkabail P, Fitzerald J, Thenkabail P (2015) Remote Sensing Handbook (Volume I): remotely sensed data characterization, classification, and accuracies. CRC Press, Boca Raton

Parajka J, Blöschl G (2008) Spatio-temporal combination of MODIS images–potential for snow cover mapping. Water Resour Res 44(3):13. https://doi.org/10.1029/2007WR006204

Parajka J, Pepe M, Rampini A, Rossi S, Blöschl G (2010) A regional snow-line method for estimating snow cover from MODIS during cloud cover. J Hydrol 381(3–4):203–212. https://doi.org/10.1016/j.jhydrol.2009.11.042

Parajka J, Haas P, Kirnbauer R, Jansa J, Blöschl G (2012) Potential of time-lapse photography of snow for hydrological purposes at the small catchment scale. Hydrol Process 26(22):3327–3337. https://doi.org/10.1002/hyp.8389

Particle (2016) IoT platform. https://www.particle.io, accessed 1 July 2016

Paudel KP, Andersen P (2011) Monitoring snow cover variability in an agropastoral area in the Trans Himalayan region of Nepal using MODIS data with improved cloud removal methodology. Remote Sens Environ 115(5):1234–1246. https://doi.org/10.1016/j.rse.2011.01.006

Pomeroy JW, Gray DM (1995) Snowcover: accumulation, relocation and management. Tech. Rep. 7, National Hydrology Research Insitute, Saskatchewan,. https://harvest.usask.ca/handle/10388/15161

Produit T (2019) Qgis plugin Pic2Map. https://github.com/tproduit/pic2map, accessed 1 July 2016

Raspberry Pi Foundation (2016) Single-board computers. https://www.raspberrypi.org, accessed 1 July 2016

Raspberry Pi Foundation (2019) Build your own weather station. https://projects.raspberrypi.org/en/projects/build-your-own-weather-station, accessed 1 October 2019

Riggs GA, Hall DK, Salomonson VV (2006) MODIS snow products user guide to collection 5. NASA/Goddard Space Flight Center, Greenbelt. https://modis-snow-ice.gsfc.nasa.gov/uploads/sug_c5.pdf

Riggs GA, Hall DK, Román MO (2016) MODIS snow products collection 6 user guide. NASA/Goddard Space Flight Center, Greenbelt. https://modis-snow-ice.gsfc.nasa.gov/uploads/C6_MODIS_Snow_User_Guide.pdf

Running SW, Mu Q, Zhao M, Moreno A (2017) MODIS global terrestrial evapotranspiration (ET) product (NASA MOD16A2/A3). Algorithm Theoretical Basis Document, Collection 5:600. https://landweb.modaps.eosdis.nasa.gov/QA_WWW/forPage/user_guide/MOD16UsersGuide2016V1.52017May23.pdf

Salomonson VV, Appel I (2004) Estimating fractional snow cover from MODIS using the normalized difference snow index. Remote Sens Environ 89(3):351–360. https://doi.org/10.1016/j.rse.2003.10.016

Sarhadi A, Kelly R, Modarres R (2014) Snow water equivalent time-series forecasting in Ontario, Canada, in link to large atmospheric circulations. Hydrol Process 28(16):4640–4653. https://doi.org/10.1002/hyp.10184

Saydi M, Ding Jl (2020) Impacts of topographic factors on regional snow cover characteristics. Water Sci Eng 13(3):171–180. https://doi.org/10.1016/j.wse.2020.09.002

Schmidt S (2009) Die reliefabhängige Schneedeckenverteilung im Hochgebirge. In: Ein multiskaler Methodenverbund am Beispiel des Lötschentals (Schweiz), Bonner Geographische Abhandlungen, vol 123. E. Ferger, Bergisch Gladbach, Germany. https://hdl.handle.net/20.500.11811/10554

Schmidt S, Weber B, Winiger M (2009) Analyses of seasonal snow disappearance in an alpine valley from micro-to meso-scale (Loetschental, Switzerland). Hydrol Process 23(7):1041–1051. https://doi.org/10.1002/hyp.7205

Sen PK (1968) Estimates of the regression coefficient based on Kendall's tau. J Am Stat Assoc 63(324):1379–1389. https://doi.org/10.1080/01621459.1968.10480934

Şensoy A, Uysal G (2012) The value of snow depletion forecasting methods towards operational snowmelt runoff estimation using MODIS and Numerical Weather Prediction Data. Water Resour Manag 26(12):3415–3440. https://doi.org/10.1007/s11269-012-0079-0

Shivhare N, Rahul AK, Dwivedi SB, Dikshit PKS (2019) ARIMA based daily weather forecasting tool: A case study for Varanasi. Mausam 70(1):133–140. https://doi.org/10.54302/mausam.v70i1.179

Sibandze P, Mhangara P, Odindi J, Kganyago M (2014) A comparison of Normalised Difference Snow Index (NDSI) and Normalised Difference Principal Component Snow Index (NDPCSI) techniques in distinguishing snow from related land cover types. S Afr J Geol 3(2):197–209. https://doi.org/10.4314/sajg.v3i2.6

Snowmetrics (2016) Snow measurement equipment and tools. https://snowmetrics.com, accessed 1 July 2016

SparkFun Electronics (2016a) Electronics retailer. https://www.sparkfun.com, accessed 1 July 2016

SparkFun Electronics (2016b) Photon weather shield hookup guide. https://learn.sparkfun.com/tutorials/photon-weather-shield-hookup-guide-v11, accessed 1 July 2016

Spell Foundry (2016) Electronic kits and accessories. https://spellfoundry.com, accessed 1 July 2016

Steiner L (2011) Reconstruction of glacier states from geo-referenced, historical postcards. Master's thesis, Institut für Kartografie und Geoinformation, Eidgenössische Technische Hochschule Zürich, Zürich. https://www.geometh-data.ethz.ch/downloads/Steiner_Masterarbeit.pdf

Stevenson TC (1866) New description of box for holding thermometers. J Scott Meteorol Soc 1:122

Stockdale CA (2017) A century of landscape change in the southern Rocky Mountains and foothills of Alberta: using historical photography to quantify ecological change. PhD thesis, University of Alberta, Department of Renewable Resources, Edmonton. https://doi.org/10.7939/R3W37M744

Stockdale CA, Bozzini C, Macdonald SE, Higgs E (2015) Extracting ecological information from oblique angle terrestrial landscape photographs: performance evaluation of the WSL Monoplotting Tool. Appl Geogr 63:315–325. https://doi.org/10.1016/j.apgeog.2015.07.012

Strausz Jr DA (2001) An application of photogrammetric techniques to the measurement of historic photographs. Master's thesis, Oregon State University, Corvallis. https://ir.library.oregonstate.edu/downloads/js956g515

Tappeiner U, Tappeiner G, Aschenwald J, Tasser E, Ostendorf B (2001) Gis-based modelling of spatial pattern of snow cover duration in an alpine area. Ecol Model 138(1–3):265–275. https://doi.org/10.1016/S0304-3800(00)00407-5

Thenkabail PS (2015) Remote sensing of water resources, disasters, and urban studies. CRC Press, Boca Raton. https://doi.org/10.1201/b19321

Tran H, Nguyen P, Ombadi M, Hsu Kl, Sorooshian S, Qing X (2019) A cloud-free MODIS snow cover dataset for the contiguous United States from 2000 to 2017. Sci Data 6:180300. https://doi.org/10.1038/sdata.2018.300

US Geological Survey (2021) Landsat Missions. https://www.usgs.gov/core-science-systems/nli/landsat, accessed 15 July 2021

Valipour M (2015) Long-term runoff study using SARIMA and ARIMA models in the United States. Meteorol Appl 22(3):592–598. https://doi.org/10.1002/met.1491

Wang J (2017) Satellite observation on snow. In: Training workshop on water resource management and sustainable development strategies in arid regions of Central/Western Asia under climate change 2017. Northwest Institute of Eco-Environment and Resources, Chinese Academy of Sciences, Lanzhou

Whittle P (1954) Hypothesis testing in time series analysis. J Am Stat Assoc 49(265):197–200. https://doi.org/10.2307/2281054

WMO (2018) Guide to climatological practices. World Meteorological Organization (WMO), Geneva. https://library.wmo.int/idurl/4/60113

Wulder MA, Masek JG, Cohen WB, Loveland TR, Woodcock CE (2012) Opening the archive: how free data has enabled the science and monitoring promise of Landsat. Remote Sens Environ 122:2–10. https://doi.org/10.1016/j.rse.2012.01.010

Xie H, Wang X, Liang T (2009) Development and assessment of combined Terra and Aqua snow cover products in Colorado Plateau, USA and northern Xinjiang, China. J Appl Remote Sens 3(1):033559. https://doi.org/10.1117/1.3265996

Yan X, Su X (2009) Linear regression analysis: theory and computing. World Scientific, Singapore. https://doi.org/10.1142/6986

Yue S, Wang C (2004) The Mann-Kendall test modified by effective sample size to detect trend in serially correlated hydrological series. Water Resour Manag 18(3):201–218. https://doi.org/10.1023/B:WARM.0000043140.61082.60

Zhang H, Zhang F, Zhang G, Che T, Yan W, Ye M, Ma N (2019) Ground-based evaluation of MODIS snow cover product V6 across China: implications for the selection of NDSI threshold. Sci Total Environ 651:2712–2726. https://doi.org/10.1016/j.scitotenv.2018.10.128

Zhang Y, Kan X, Ren W, Cao T, Tian W, Wang J (2017) Snow cover monitoring in Qinghai-Tibetan Plateau based on Chinese Fengyun-3/VIRR data. J Indian Soc Remote Sens 45(2):271–283. https://doi.org/10.1007/s12524-015-0527-4

Chapter 3
Snow Cover Distribution in Ladakh

Abstract The analyses of the snow cover distribution in Ladakh are the main focus of this chapter, and they traverse diverse temporal scales, encompassing monthly, seasonal, and annual dimensions, all the while maintaining a pixel-wise granularity at both regional and subregional levels. The work also showcases the synergistic potential of leveraging modern terrestrial camera technology to complement remote sensing data across diverse topographies. Focusing on MODIS snow cover products, it examines snow cover area (SCA) patterns and pixel-precise snow cover days (SCD) definitions. Monthly SCA trends over nearly two decades in Ladakh are visually presented. Statistical tests reveal positive trends in summer and year-round, and negative trends in autumn, winter, and spring, although not significant. The study assesses the topography's influence on snow cover distribution across seven subregions based on altitude, slope inclination, and orientation. An Autoregressive Integrated Moving Average (ARIMA) model predicts snow cover dynamics, comparing measured and modelled data. Field-collected snow thickness and water equivalent data validate findings. The snow line for the winter of 2016 presents compelling results. This research provides insights into Ladakh's snow dynamics, bridging remote sensing and ground observations, and highlighting topographical impacts.

Keywords Snow · Cloud reduction · Pixel-wise analysis · Topographical impact · Time series · ARIMA model · Snow stake · Ladakh · India

3.1 Reducing MODIS Cloud Obscuration and the Spatio-Temporal Composite

The monthly mean percentages Moderate Resolution Imaging Spectroradiometer (MODIS) cloud cover area (CCA), including the reduction of the cloud cover in three steps, are illustrated in Fig. 3.1. From a global perspective, the CCA over land is slightly higher for MODIS Aqua than for Terra (Riggs et al. 2006).

S. Passang, *Snow Cover Distribution and Dynamics*, Advances in Asian Human-Environmental Research, https://doi.org/10.1007/978-3-031-57692-8_3

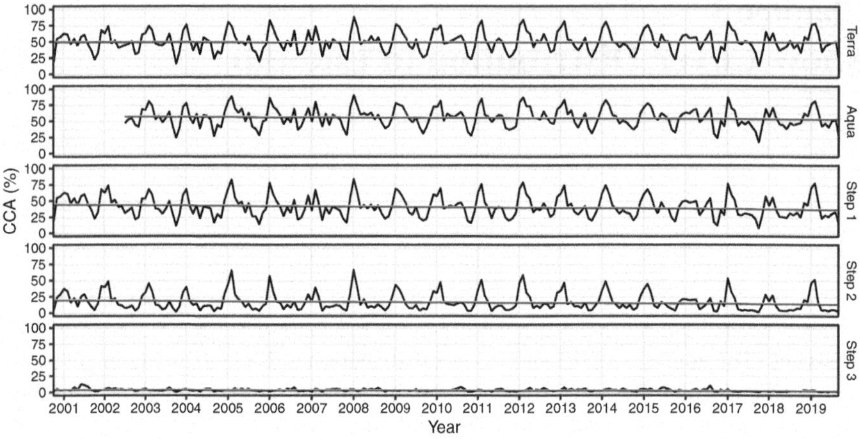

Fig. 3.1 Monthly average percentages of the CCA of MODIS daily snow product data featuring the reduction of cloud coverage (in a black line) in three steps. The overall mean percentages are shown in a blue line. On average, the Terra and Aqua cloud coverage amount to, respectively, 49.54% and 55.36%, and the smallest Terra and Aqua cloud coverage was observed to be, respectively, 11.80% and 17.19% in October 2017. Whereas in January 2008 87.86% of the study area was cloud-covered for Terra and 90.29% for Aqua

The first cloud reduction step, based on the combination of Terra and Aqua MODIS products, reduced the monthly average percentages of the CCA to a mean of 41.15%, with a minimum cloud coverage of 7.41% and maximum cloud coverage of 83.60%. The second cloud reduction step, derived from the temporal combination decreased the monthly average percentages of CCA to 16.66%, whereas the minimum cloud coverage amounted to 0.82% and the maximum cloud coverage amounted to 65.94%. The remaining 3.12% mean, minimum of 0% and maximum of 18% CCA, between the Regional Snow Line Elevation (RSLE) and Regional Land Line Elevation (RLLE), was classified as mixed pixels, resulting in the final spatio-temporal composite of cloudless snow maps, facilitating a better monitoring of the snow cover distribution in the study area.

The final output MODIS cloudless snow cover product was categorised into land, snow, and mixed raster image pixels. Figure 3.2 shows daily percentages of the land cover area (LCA), snow cover area (SCA), and mixed cover area (MCA) of the MODIS cloudless spatio-temporal composite, in the period from October 2000 until September 2019 for the study area.

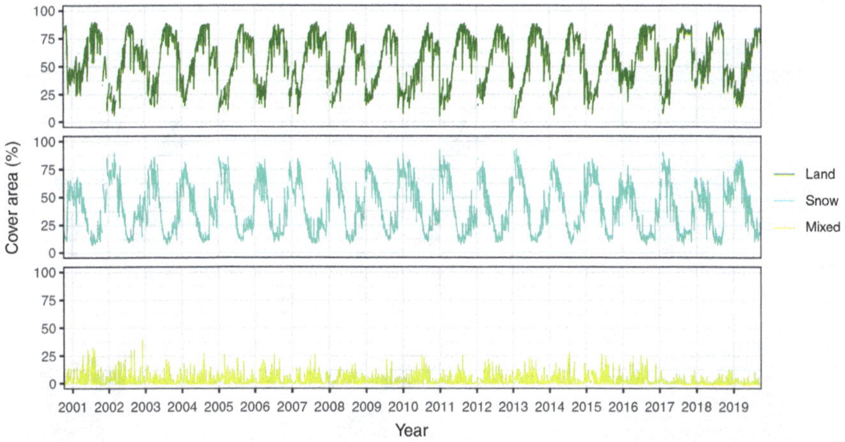

Fig. 3.2 Daily percentages of the land, snow, and mixed cover areas of the MODIS spatio-temporal combination. The daily percentages of the LCA show a minimum at 3.35%, first quarter at 37.73%, median at 58.78%, mean at 56.14%, third quarter at 76.40%, and maximum at 91.44%. The daily percentages of the SCA show a minimum at 6.45%, first quarter at 19.88%, median at 38.00%, mean at 40.77%, third quarter at 59.15%, and maximum at 93.34%. The daily percentages of the MCA show a minimum at 0%, first quarter at 0.34%, median at 1.47%, mean at 3.09%, third quarter at 4.06%, and maximum at 40.10%

3.2 Validation and Accuracy Assessment

To assess the reliability of the composite cloudless MODIS snow cover maps, cloud-free Landsat 8 Operational Land Imager (LC8 OLI) images were used to validate MODIS images with the Cohen's kappa coefficient (Cohen 1960) method, on a regional scale. In a second approach on a local scale, Landsat 8 OLI and MODIS snow cover maps were compared to the orthorectified terrestrial images taken on the same date.

3.2.1 Regional Scale

The comparison during four different seasons of 16 MODIS Terra and Aqua satellite Normalised Difference Snow Index (NDSI) snow cover maps with different ranges of NDSI values, with 8 Landsat 8 OLI satellite "true" snow cover maps, divided into a northern and southern tile, using the kappa coefficient method, shows that the relevant NDSI threshold value for the Trans-Himalayan region of Ladakh is 0.40.

Figure 3.3 illustrates an example of Landsat 8 tiles against MODIS Terra NDSI snow cover during the winter. The original Landsat images are shown in cyan for snow and brown for no snow, the Landsat "true" snow and MODIS Terra images

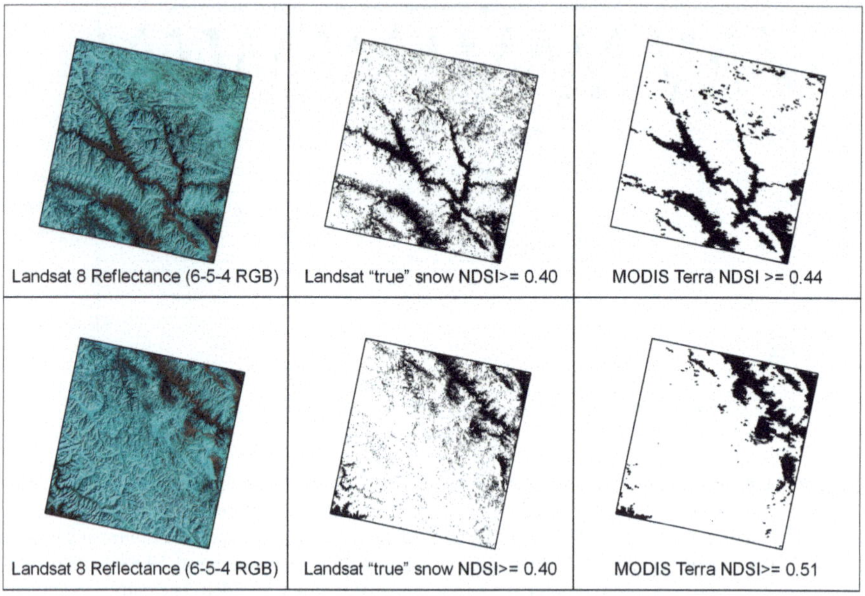

Fig. 3.3 MODIS NDSI snow cover against Landsat 8 OLI "true" snow in winter (Top row, northern tile; and bottom row, southern tile images of 24 February 2017)

with the NDSI values resulting from the highest kappa coefficient, are shown in white for snow and black for no snow.

The results of the tests show that there exists spatial and seasonal variation of the NDSI threshold value in the study area. Especially in winter and spring the NDSI predominantly exceeds the reference threshold value of 0.40, but summer and autumn show lower values again. Table 3.1 gives a summary of the tested 16 MODIS against 8 Landsat satellite tiles, the different seasons and resulting NDSI threshold values with their average and overall accuracy. For a complete overview of all NDSI threshold results see Table A.1.

3.2.2 Local Scale

In the validation care was taken to use images of both MODIS collection 5 (images of 2016) and 6 (images of 2017), to enhance the comparison between the two versions. Figure 3.4 presents the validation of snow cover at the Gyamtsa and Lungmar locations, and Fig. 3.5 shows the validation at Khakshal and South Phulu.

The results are well in congruence with the orthorectified photos. Comparison amongst the coarse resolution MODIS, high Landsat and very high-resolution terrestrial orthorectified images, shows that the resolution of pixels plays a huge role in detecting shallow and fragmented snow traces in the transition zone. MODIS

Table 3.1 NDSI average of 16 MODIS against 8 Landsat satellite tiles. The average value and overall accuracy are highlighted in red

MODIS satellite	Tile	Season	NDSI	Accuracy (%)
Terra	North	Spring	0.39	87.43
Terra	South	Spring	0.46	85.22
Aqua	North	Spring	0.34	86.68
Aqua	South	Spring	0.42	85.59
Terra	North	Summer	0.39	91.26
Terra	South	Summer	0.34	92.36
Aqua	North	Summer	0.32	91.94
Aqua	South	Summer	0.33	93.49
Terra	North	Autumn	0.37	83.55
Terra	South	Autumn	0.45	81.11
Aqua	North	Autumn	0.36	83.15
Aqua	South	Autumn	0.45	82.57
Terra	North	Winter	0.44	87.42
Terra	South	Winter	0.49	90.78
Aqua	North	Winter	0.43	85.66
Aqua	South	Winter	0.49	86.93
	NDSI	**Average**	**0.40**	**87.20**

underestimates the registration of snow pixels where a pixel is less than 50% snow covered and overestimates where a pixel is more than 50% snow covered (Hall and Riggs 2007). For example, ephemeral shallow and fragmented snow pixels in the near range of the Lungmar camera are not registered by MODIS (Fig. 3.4i) but detected by Landsat (Fig. 3.4h). However, MODIS in the permanent snow cover regions overestimates a bit where a pixel is more than 50% snow covered (Fig. 3.4l), in comparison with Landsat (Fig. 3.4k).

3.3 Trend Analysis of Snow Cover Area and Snow Cover Days

The monthly average snow cover percentages for the entire study area are presented in Fig. 3.6. The snow accumulation process initiates from September, reaches its peak in February–March, and gradually ablates throughout the spring and summer months, to reach its minimum in August. July revealed a statistically significant positive trend of the SCA of 0.3% year^{-1}. The mean shows a minimum in summer (19.90%) and maximum in winter (61.17%), and the standard deviation (SD) a minimum in summer (3.95%) and maximum in autumn (6.50%). The annual mean and SD are 41.85% and 4.32%, respectively. A slight negative trend of the SCA is observed in autumn, winter, and spring, while a slight positive trend of SCA is observed annually (0.08 year^{-1}) and in summer (0.2% year^{-1}). However, these

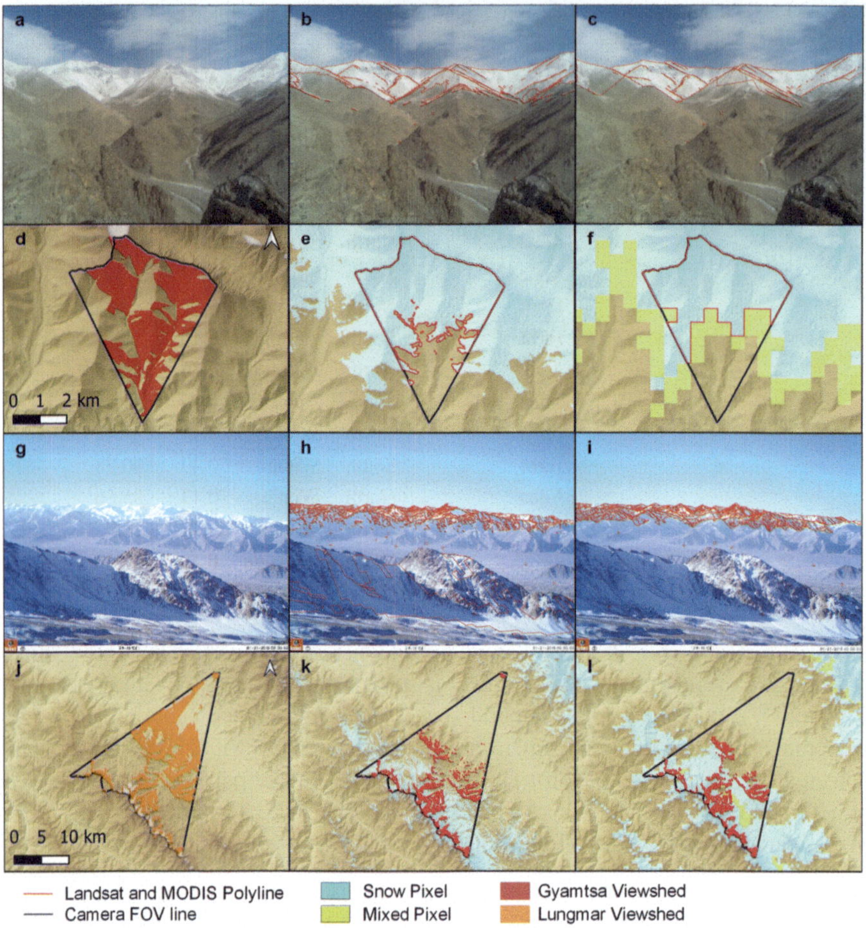

Fig. 3.4 Oblique terrestrial image taken from Gyamtsa on 23 January 2017 (**a**); Orthorectified and overlaid with snow polygons (red lines) derived from Landsat 8 OLI image (**b**) and MODIS (**c**); Gyamtsa camera viewshed in red colour (**d**); Landsat 8 OLI (**e**) and MODIS snow shown in cyan (**f**); Oblique terrestrial image taken from Lungmar on 21 January 2016 (**g**); Orthorectified and overlaid with snow polygons (red lines) derived from Landsat 8 OLI image (**h**) and MODIS (**i**); Lungmar camera viewshed in orange colour (**j**); Landsat 8 OLI (**k**) and MODIS snow (**l**) (Some photos show control points with plus and multiply signs in red)

trends are not found to be statistically significant. See Fig. 3.7, Tables 3.2 and D.1 for more details.

The trend analysis of SCA on a subregional scale shows slight increasing significant trends in summer in the regions of Suru (0.52% year^{-1}) and Zangskar (0.28% year^{-1}), with p-values of 0.05 and 0.09, respectively, while in Indus, Pangong, and Shayok a minor insignificant decreasing trend of SCA can be detected

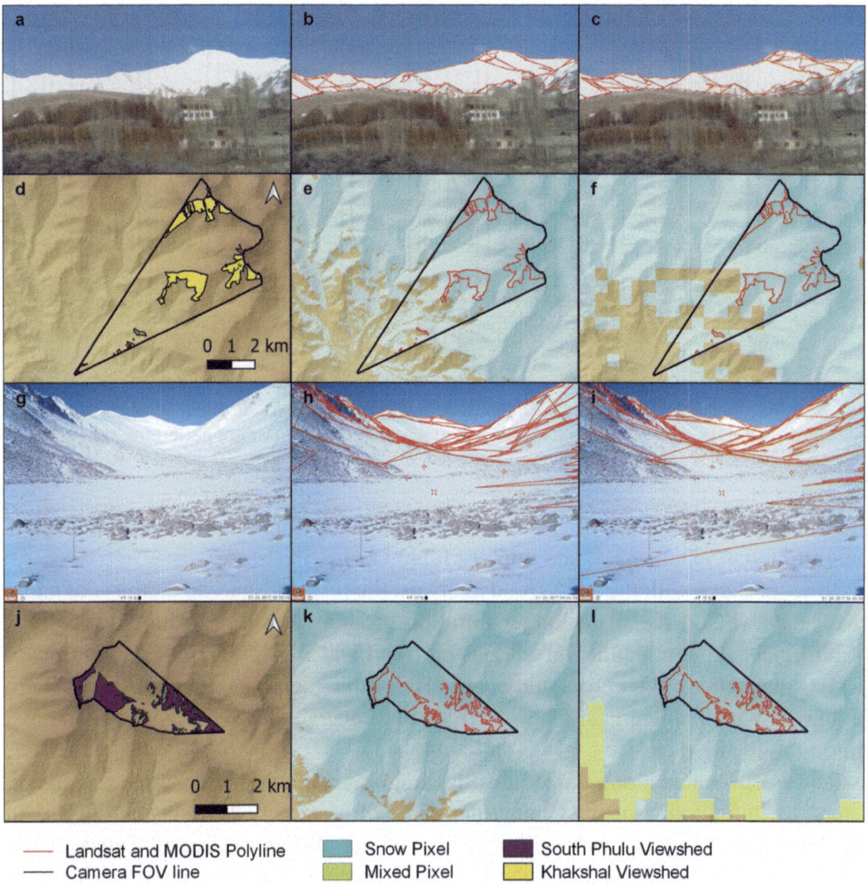

Fig. 3.5 Oblique terrestrial image taken from Khakshal on 22 February 2017 (**a**); Orthorectified and overlaid with snow polygons (red lines) derived from Landsat 8 OLI image (**b**) and MODIS (**c**); Khakshal camera viewshed in yellow colour (**d**); Landsat 8 OLI (**e**) and MODIS snow shown in cyan (**f**); Oblique terrestrial image taken from South Phulu on 23 January 2017 (**g**); Orthorectified and overlaid with snow polygons (red lines) derived from Landsat 8 OLI image (**h**) and MODIS (**i**); South Phulu camera viewshed in purple colour (**j**); Landsat 8 OLI (**k**) and MODIS snow (**l**) (Some photos show control points with plus and multiply signs in red)

in winter and spring. Siachan and Tshomoriri show a subtle (non-significant) negative trend in winter (see also Table D.3).

Determined by the SCA, RSLE features an ascending and descending pattern in the study region. It ascends from 4813 m a.s.l. in February to 5097 m a.s.l. in April, with a maximum of 5560 m a.s.l. in August, from where it descends again. The monthly, seasonal and annual average RSLE shows negative trends in the months February–November, autumn, spring, summer and annually. However, only the downward shifts of the RSLE in June (–5.20 m year^{-1}), July (–5.56 m year^{-1}),

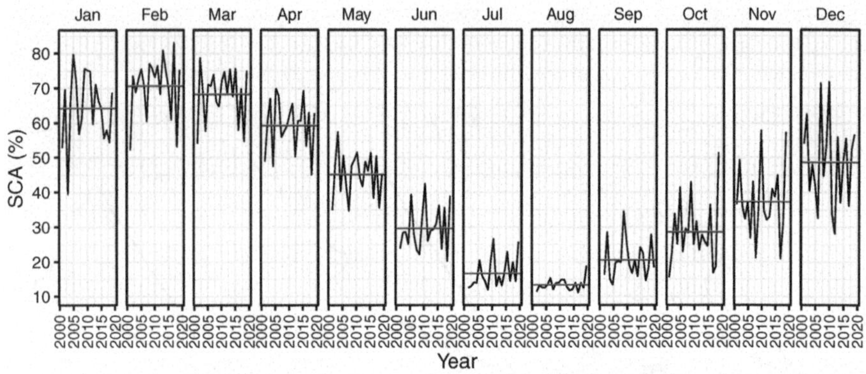

Fig. 3.6 Monthly mean percentages (in a black line), with the overall monthly mean (in a blue line) of the MODIS SCA in the study area of Ladakh over the observation period of 2000–2019

and summer (–3.53 m year^{-1}) are significant. Positive non-significant trends show only in January, December, and winter (Table D.2).

In accordance with the SCA trend analysis, the pixel-wise trend analysis of snow cover days (SCD) reveals a longer duration of snow coverage. Annually just 6% of all pixels show a significant increasing trend and 2% of all pixels a significant decreasing trend in the entire study area. The results demonstrate that the SCD are highest in winter and almost identical in spring, relatively lower in autumn, and at a minimum in summer (see Fig. 3.8 and Table 3.3).

The areas located along the large valley bottoms of the Indus, Nubra, and Shayok rivers, including parts of the Changthang area of eastern Ladakh, remain less than 10 days completely snow-free during the winter. Whereas areas located in the subregion of Siachan and along the Greater Himalayan range remain permanently snow covered (see Fig. 3.8, top).

The average SCD of the entire study area is 151 days, and the average SCD amongst the subregions bears out a minimum for Indus (87 days) and maximum for Siachan (235 days). Averages of SCD in the rest of the subregions in a decreasing order reveal: Suru (184 days), Zangskar (161 days), Shayok (157 days), Tshomoriri (126 days), and Pangong (105 days).

Figure 3.8, bottom, shows that subregions with a significant increasing pixel trend at p-value < 0.1 are located in Suru (36%), Zangskar (18%), Tshomoriri (10%), Siachan (9%), Shayok (6%), Pangong (5%), and Indus (4%) during the summer, whereas the subregion of Siachan has a significant decreasing pixel trend of 4%. In the winter there are significant decreasing trends mainly in the eastern parts of Ladakh, e.g., in Pangong (5%), Shayok (3%), Indus (2%), and Tshomoriri (2%). During spring spread-out portions with significant increasing pixel trends are noticeable in the Suru (14%) and Zangskar (4%) areas, and chunks of significant decreasing pixel trend are concentrated in the eastern parts of the Shayok and Indus regions (both 2%), with here and there scattered heterogeneous

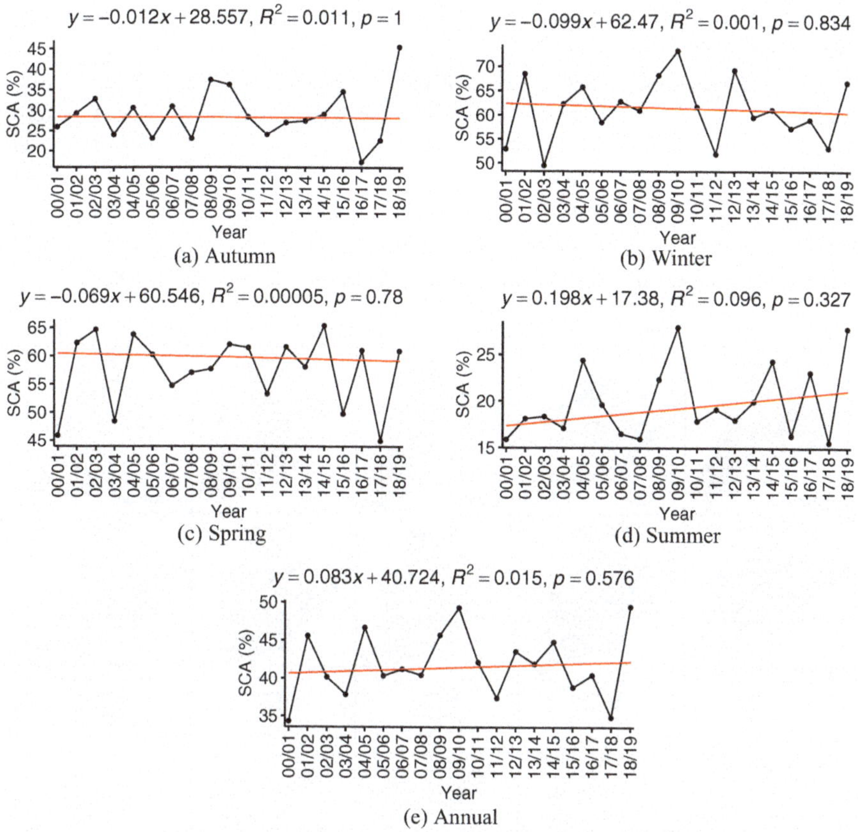

Fig. 3.7 Seasonal and annual Mann–Kendall test statistics of the mean MODIS SCA percentages (black line), with the magnitude of the Sen's slope line (red line). Years are shown in hydrological year format, with 00/01 meaning 2000/2001, etc. (Study area of Ladakh over the period of 2000–2019)

significant trends in the rest of the subregions. In the autumn season the increasing and decreasing significant pixel trends are at a minimum in all regions, with the exception of Siachan, Suru, and Shayok, that show, respectively, 5%, 4%, and 3% of increasing trend. Annually, it is observed that the increasing pixel trends are mainly concentrated in Suru (25%) and Zangskar (7%), while the highest decreasing trend is in the Siachan region (5%).

Table 3.2 Seasonal and annual Mann–Kendall test statistics of the average SCA (%) in the study area of Ladakh in the period from 2000–2019

	Linear slope	Sen's slope	Z-value	*p*-value	Trend	Mean	SD
Autumn	0.12	−0.01	0	1.00	▼	28.93	6.50
Winter	0.03	−0.10	−0.21	0.83	▼	61.17	6.49
Spring	−0.01	−0.07	−0.28	0.78	▼	57.53	6.33
Summer	0.22	0.20	0.98	0.33	▲	19.90	3.95
Annual	0.09	0.08	0.56	0.58	▲	41.85	4.32

SD standard deviation, *filled triangle* positive trend, *filled inverted triangle* negative trend

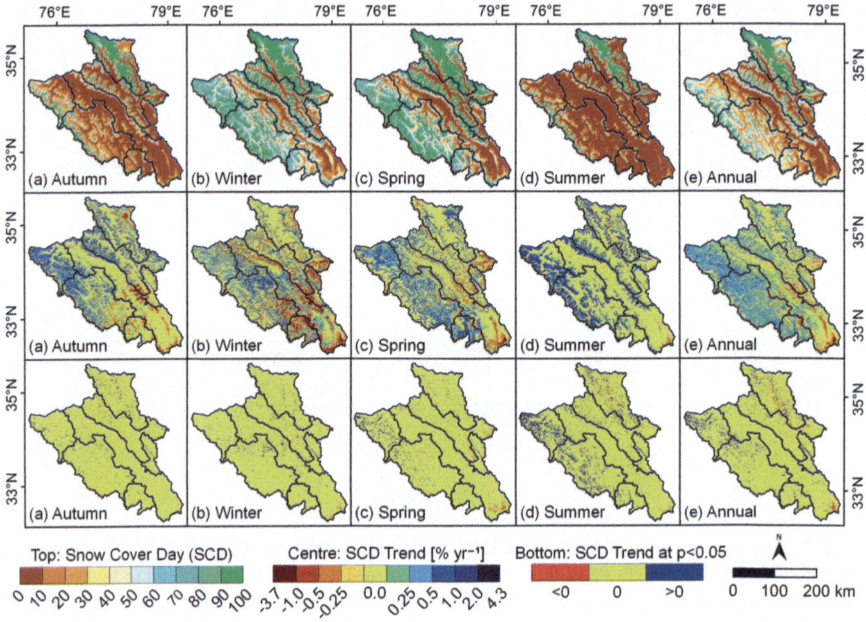

Fig. 3.8 Pixel-wise analysis of snow cover day (SCD) for the entire study area of Ladakh in the observation period 2000–2019. Subregions are demarcated in black lines. Top: mean seasonal and annual SCD; centre: seasonal and annual trends of SCD; bottom: trend significance of SCD, where yellow colour symbolises no significant trend, red negative and blue positive trends, with p-value < 0.1

3.4 Topographical Effects on Snow Cover Distribution

The results of the impact of topographic (elevation, slope, and aspect) effects on the MODIS SCA in the seven subregions, in order to check the spatio-temporal variability in the study region over a period of almost two decades (2000–2019), are presented below.

Table 3.3 Increasing and decreasing trend of SCD in the study area of Ladakh for the observation period of 2000–2019

Increasing trend (Sen's slope > 0)								
	All		$p < 0.01$		$p < 0.05$		$p < 0.1$	
	km^2	%	km^2	%	km^2	%	km^2	%
Autumn	23538.50	35.83	126.00	0.19	710.00	1.08	1571.00	2.39
Winter	21587.75	32.86	103.75	0.16	559.50	0.85	1280.75	1.95
Spring	25163.25	38.31	205.75	0.31	1011.00	1.54	1993.00	3.03
Summer	24815.00	37.78	725.75	1.10	4269.00	6.50	7853.75	11.96
Annual	33409.50	50.86	346.50	0.53	1934.75	2.95	4149.25	6.32
Decreasing trend (Sen's slope < 0)								
	All		$p < 0.01$		$p < 0.05$		$p < 0.1$	
	km^2	%	km^2	%	km^2	%	km^2	%
Autumn	9907.75	15.08	11.75	0.02	48.75	0.07	136.50	0.21
Winter	20396.00	31.05	36.50	0.06	399.75	0.61	1027.25	1.56
Spring	12440.50	18.94	47.75	0.07	365.00	0.56	811.25	1.23
Summer	3777.25	5.75	152.25	0.23	447.00	0.68	731.50	1.11
Annual	11983.25	18.24	230.50	0.35	743.25	1.13	1220.00	1.86

p significance level

3.4.1 SCA in Relation to Subregional Elevation Zones

The SCA increases with elevation, where areas above 6500 m a.s.l. are mostly covered in snow due to the perennial snow. Seasonal variation in the subregions of Siachan and Suru is minimal, as more than 80% of the elevation zone of 5500–6500 m a.s.l. is snow covered, whereas seasonal variation fluctuating from 29% in summer to 70% in winter in other subregions is observed. In spring the lower elevation classes below 4500 m a.s.l. start ablating while above 4500 m a.s.l. there is still accumulation in all the subregions, as the higher elevations receive precipitation in the form of snow throughout all the seasons.

The SCA varies between 15% in summer and 64% in winter for the elevation zone of 4500–5500 m a.s.l., and below 4500 m a.s.l. snow cover is detected in all subregions except Pangong, as the SCA is less than 40% in winter. The winter snow cover is less than 20% in the northern subregions, while it reaches up to 60% in Suru and 40% in Zangskar for the elevation zone below 3500 m a.s.l.

The trend analysis of the SCA in relation to elevation shows a heterogeneous pattern. In all subregions, for the elevation zones below 3500 m a.s.l. and 3500–4500 m a.s.l., the increasing trends are mostly significant, whereas slight significant decreasing trends above 5500 m a.s.l. and mostly significant decreasing trends above 6500 m a.s.l. can be detected. Subtle significant decreasing trends in spring and mostly significant increasing trends in summer are observed for elevation zone 4500–5500 m a.s.l. in all subregions. For example, elevations above 6500 m have a highly significant decreasing trend in winter (–0.10% year^{-1}) and in summer (–

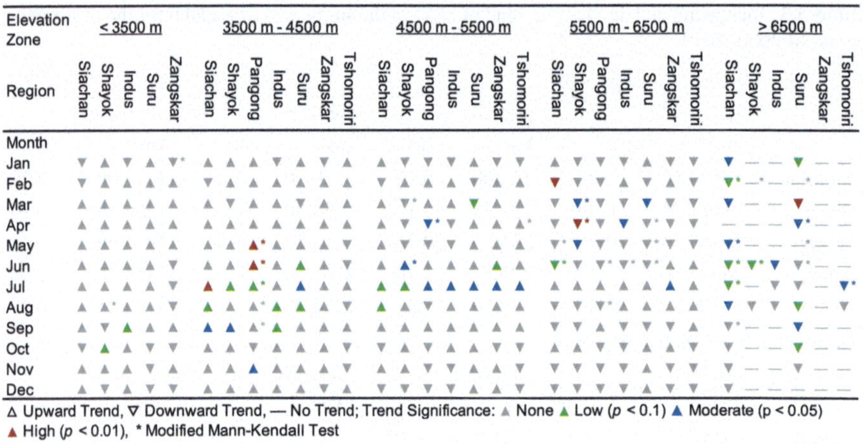

Fig. 3.9 Monthly SCA trend in the elevation zones of the seven subregions in the study area of Ladakh for the observation period of 2000–2019

0.26% year^{-1}) in the subregions of Suru and Tshomoriri, respectively. On the other hand there is a highly significant increasing trend for elevation zone 3500–4500 m in summer (0.08% year^{-1}) in the subregion of Pangong. See also Figs. 3.9 and 3.10.

3.4.2 SCA in Relation to Subregional Slope Zones

A decrease of SCA with an increasing slope angle is detected in the Shayok region, whereas in other subregions the SCA does not show large differences. SCA in mid-range slopes from 10°–30° is detected in a higher proportion in all subregions. In comparison to other subregions, Indus has the lowest amount of SCA in all slope angles. The Pangong and Tshomoriri slope gradient above 40° is missing due to gradual slopes in these higher plateau regions.

The trend analysis of the SCA in relation to slope displays a significant decreasing trend for the slope zone of 30°–40° in winter (−0.65% year^{-1}) in the subregion of Pangong. Whereas a significant increasing trend for the slope zone above 40° in autumn (0.03% year^{-1}) and winter (0.04% year^{-1}) is observed in subregion Indus.

A highly significant increasing trend can be detected in summer for the slope zones below 10° and 20–30° (with 0.5% and 0.05% year^{-1}, respectively) in the subregion of Suru, and in Zangskar (0.2% and 0.32% year^{-1}). However subregions Siachan, Shayok, and Tshomoriri show mostly non-significant heterogeneous patterns during all seasons. See also Figs. 3.11 and 3.12.

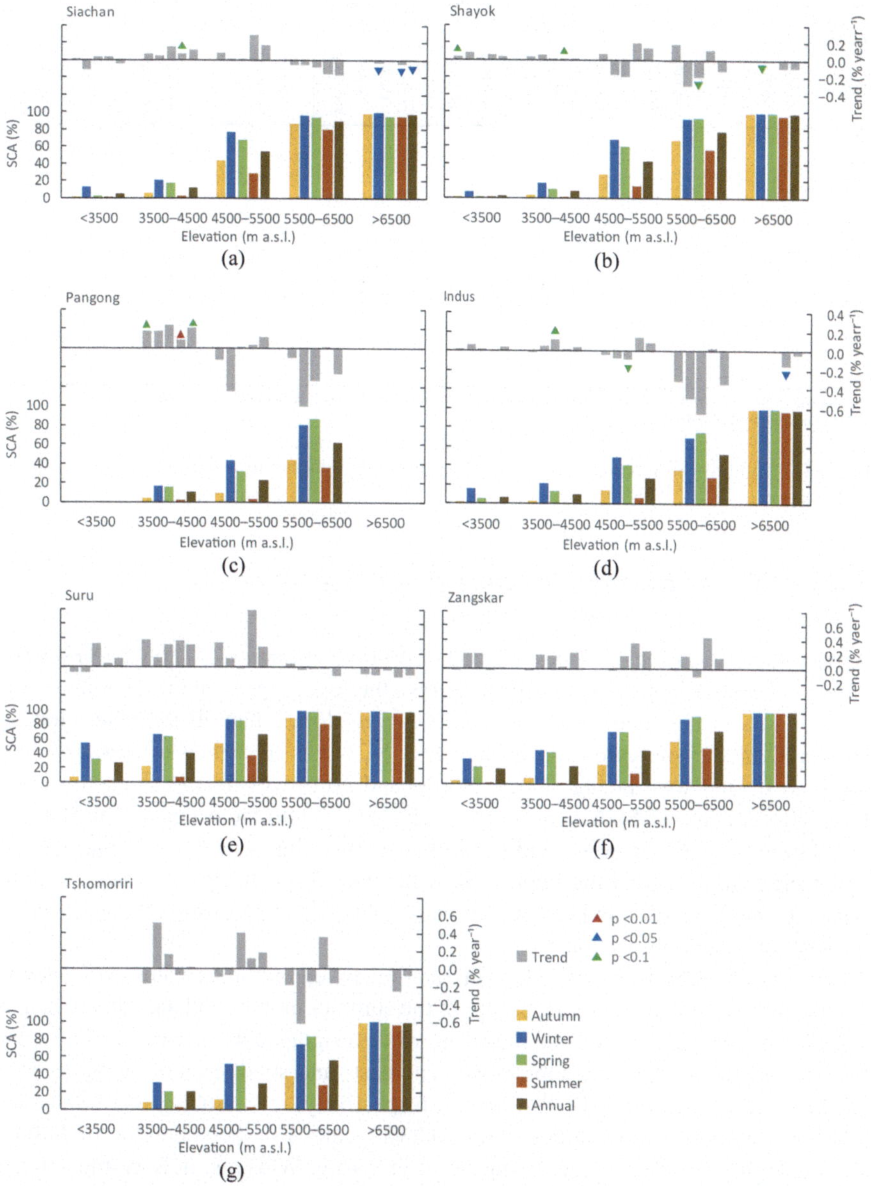

Fig. 3.10 Seasonal and annual MODIS SCA and trends in relation to the elevation zones of the subregions in Ladakh. Three p-value significance levels are shown, indicated with red ($p < 0.01$); blue ($p < 0.05$); and green ($p < 0.1$) triangle symbols

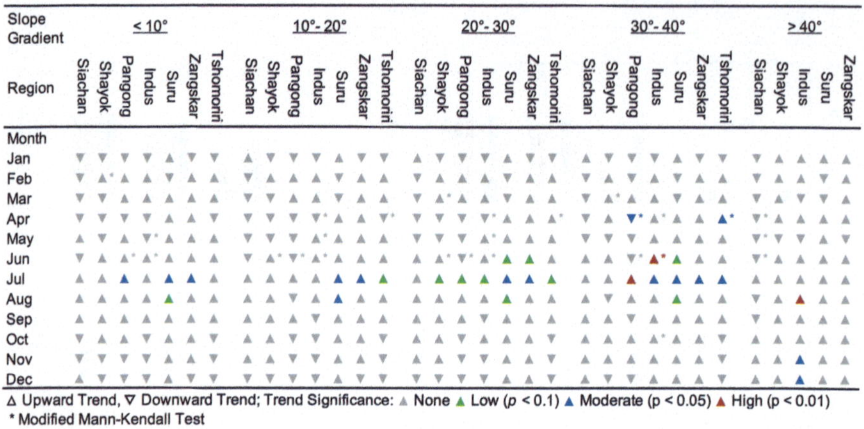

Fig. 3.11 Monthly SCA trend in five slope categories of the subregions in the study area of Ladakh in the period of 2000–2019

3.4.3 SCA in Relation to Subregional Aspect Zones

Aspect, the orientation of slopes, is another prominent feature of snow cover distribution. The SCA in relation to aspect follows the topography, and accumulation of snow is less in the south- and south-west-facing slopes than all the other slopes. However, in the subregion of Siachan less SCA has a slight shift towards the south-west- and west-facing slopes. North- and north-west-facing slopes observe the highest degree of SCA in almost all subregions, with the exception of Shayok and Tshomoriri, where west- and north-west-facing slopes, and north- and north-east-facing slopes detect the highest SCA amount, respectively. Moreover, a high seasonal characteristic can be observed in all subregions apart from Siachan, where it shows only subtle variation.

The trend analysis of the SCA in relation to eight aspect categories shows diverse non-significant patterns, except in the summer months and annually, when a significant increase in trend, and in spring, when a significant decrease in trend, can be detected for most of the subregions. The east- and south-eastern-facing slopes (with –0.93% year^{-1} and –0.10% year^{-1}, respectively) in the subregion of Shayok, and north-western-facing slopes (with –0.23% year^{-1}) in the subregion of Indus, show a highly significant decreasing trend in spring. Whereas, in Suru the south-western-facing slopes have a highly significant increasing trend in summer, with 0.91% year^{-1}. See also Figs. 3.13 and 3.14.

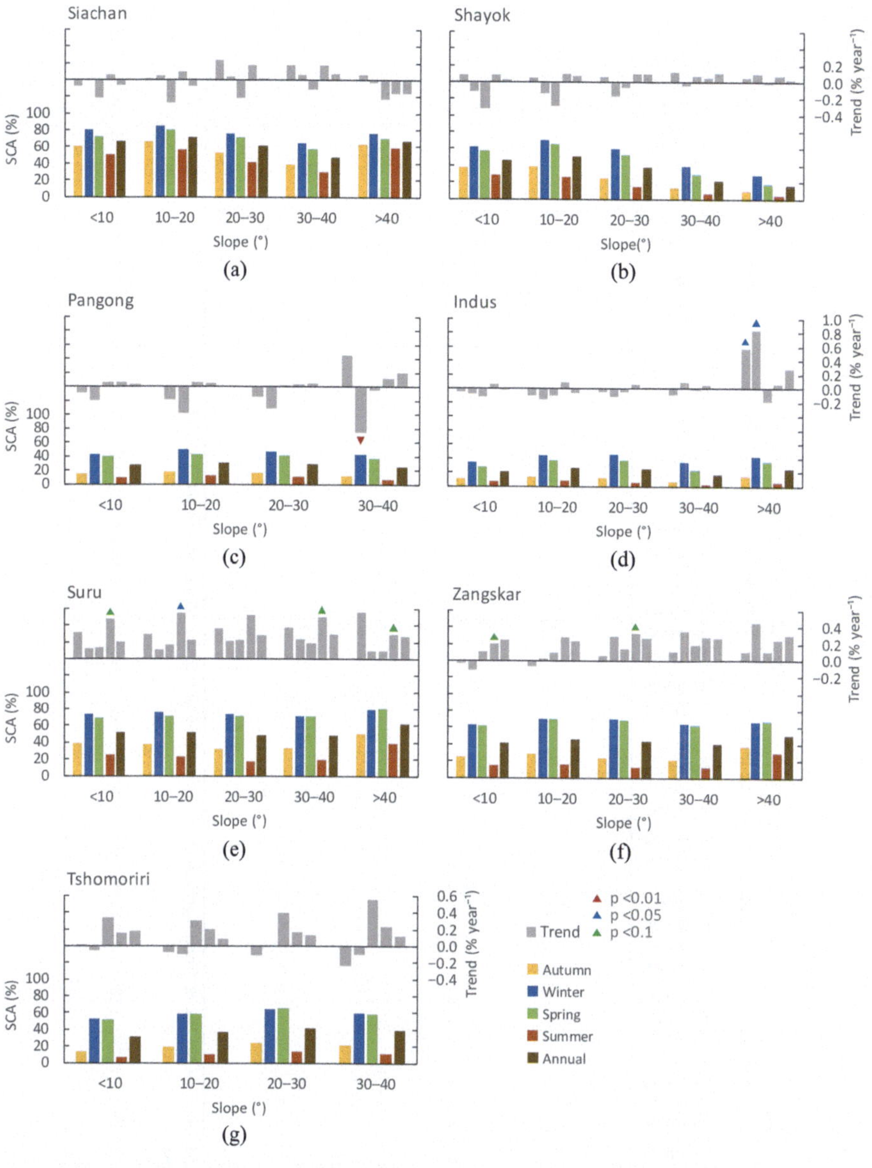

Fig. 3.12 Seasonal and annual MODIS SCA and trends in relation to the slope zones of the subregions in the study area of Ladakh over the observation period 2000–2019

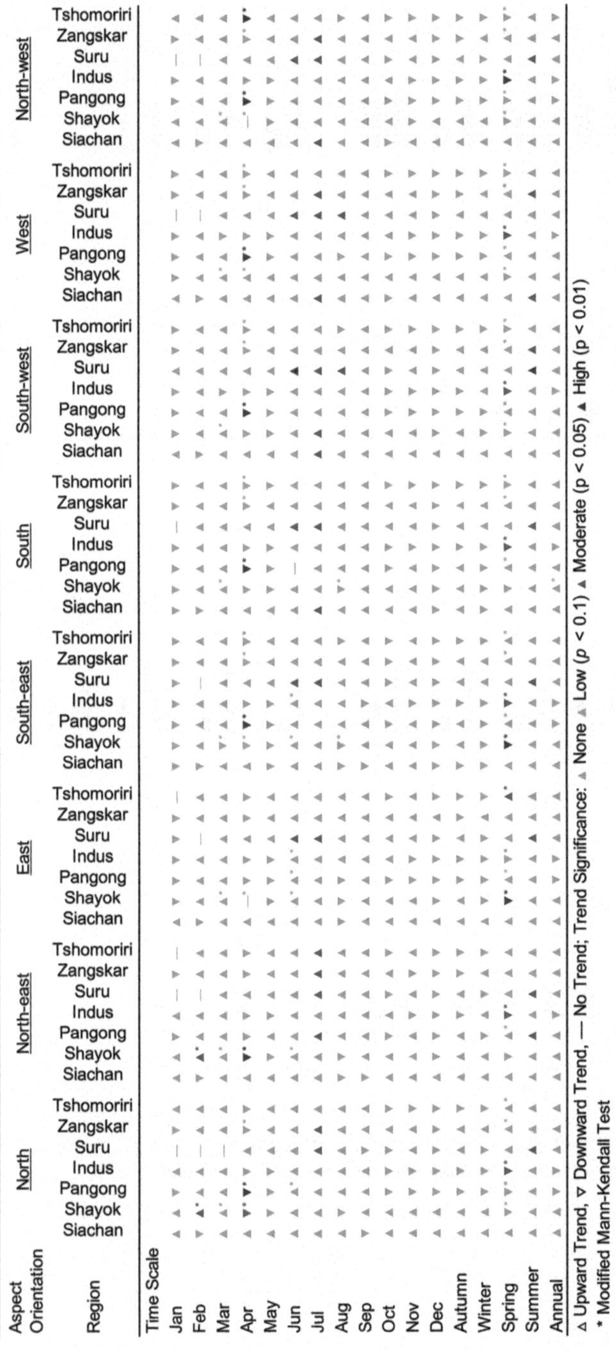

Fig. 3.13 Monthly, seasonal, and annual SCA trend in eight aspect orientations of the subregions in the study area of Ladakh over the period of 2000–2019

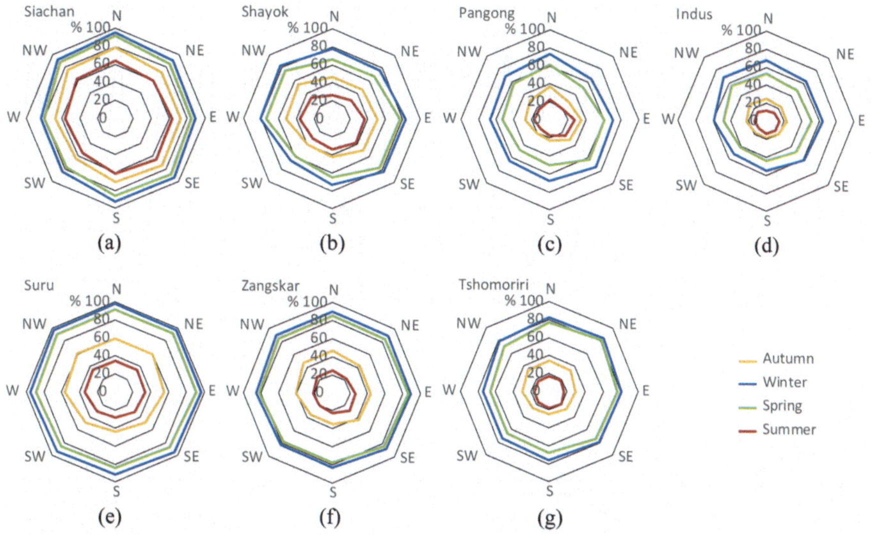

Fig. 3.14 Seasonal and annual MODIS SCA in connection to the aspect zones of the subregions in the study area of Ladakh from 2000–2019

3.5 Snowfall Outlook Based on an ARIMA Model

3.5.1 Exploratory Data Analysis

Figure 3.15a forms the core time series in this analysis. The graph features the MODIS SCA in the study area, on a yearly basis calculated from monthly mean percentages, starting from 1 October 2000 until 30 September 2019 (19 full hydrological snow cycles). Figure 3.15b features a box plot of the same. For a list of the MODIS SCA monthly areal proportions and areal extents, see Table 3.4.

From these exploratory data, we can make some initial inferences: The yearly fluctuations of the snowfall time series show a seasonal and fairly constant size pattern, with an apparent cycle of 12 months.

Clearly the winter months December-February have the most snowfall, with higher means and variance than the other months. The spring months March-May have more snowfall with higher means and variance than the autumn months September-November. The summer months July-August have snowfall with the least means and variance.

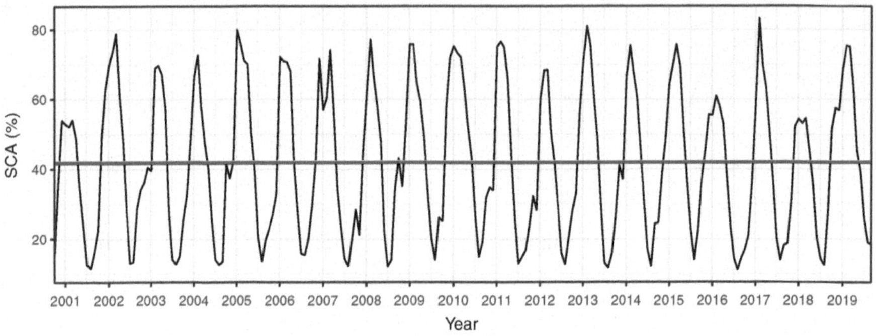

(a)

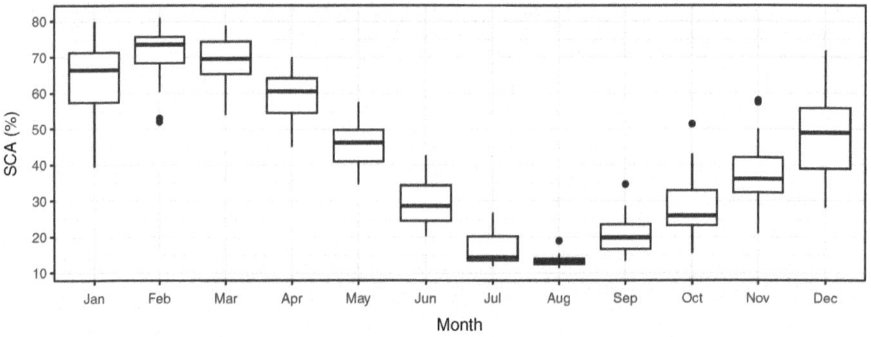

(b)

Fig. 3.15 Yearly monthly mean percentages of the MODIS SCA. (**a**) Time series, with the average mean line (at 41.85%) shown in blue; and (**b**) Box plot (Study area of Ladakh over the period 2000–2019)

3.5.2 Time Series Analysis and Validation

In this study an additive decomposition of the time series was chosen, because the seasonal fluctuations of the time series are quite constant in size over time (Cryer and Chan 2008; Everitt and Skrondal 2010; Chatfield and Xing 2019; Hyndman and Athanasopoulos 2021) (see Fig. 3.15a).

Figure 3.16 shows the graphical result of this decomposition. The three decomposed components are shown separately in the bottom three panels. These components can be added together to reconstruct the original time series data, shown in the top panel (same as Fig. 3.15a). The seasonal values' component clearly show that there is a repeating cycle of snowfall every year. The trend values show that there is an alternating trend of increase and decrease in the values of the time series. The remainder values show the irregularity in the values of the time series, left over when the seasonal and trend components have been subtracted from the data. They do not show a fixed frequency.

Table 3.4 MODIS SCA yearly monthly areal extents (AE, in km^2) and areal proportions (AP, in %) in the study area of Ladakh from 2000–2019

Year		Jan	Feb	Mar	Apr	May	Jun	Jul	Aug	Sep	Oct	Nov	Dec
2000	AE										10150	23936	35404
	AP										15.45	36.44	53.90
2001	AE	34676	34177	35470	32053	22847	15651	8182	7429	10696	14729	32351	41064
	AP	52.79	52.03	54.00	48.80	34.78	23.83	12.46	11.31	16.28	22.42	49.25	62.51
2002	AE	45681	48263	51700	39698	31383	18602	8434	8635	18720	22242	23591	26484
	AP	69.54	73.47	78.70	60.43	47.77	28.32	12.84	13.14	28.50	33.86	35.91	40.32
2003	AE	25879	45163	45675	43992	37711	18769	9163	8190	9758	16469	21156	31510
	AP	39.40	68.75	69.53	66.97	57.41	28.57	13.95	12.47	14.86	25.07	32.21	47.97
2004	AE	43626	47648	37857	31097	26399	16415	9075	8154	8747	27172	24414	27577
	AP	66.41	72.54	57.63	47.34	40.19	24.99	13.81	12.41	13.32	41.36	37.17	41.98
2005	AE	52405	49604	46637	45885	33178	25823	13454	8773	12953	15006	17533	21317
	AP	79.78	75.51	71.00	69.85	50.51	39.31	20.48	13.36	19.72	22.84	26.69	32.45
2006	AE	47277	46463	46328	44537	27777	18270	10334	10003	13262	19402	28287	46903
	AP	71.97	70.73	70.53	67.80	42.29	27.81	15.73	15.23	20.19	29.54	43.06	71.40
2007	AE	37257	39631	48542	36618	22739	15370	9293	7818	13055	18552	13839	29201
	AP	56.72	60.33	73.90	55.75	34.62	23.40	14.15	11.90	19.87	28.24	21.07	44.45
2008	AE	40160	50591	43333	37808	31299	14529	7784	9078	22659	28205	22896	35008
	AP	61.14	77.02	65.97	57.56	47.65	22.12	11.85	13.82	34.49	42.94	34.85	53.29
2009	AE	49667	49687	42477	38880	32324	21624	13298	9105	17051	16388	37999	47153
	AP	75.61	75.64	64.66	59.19	49.21	32.92	20.24	13.86	25.96	24.95	57.85	71.78

(continued)

Table 3.4 (continued)

2010	AE	49307	47954	47216	41134	33796	27898	17457	9684	12518	20794	22736	22192
	AP	75.06	73.00	71.88	62.62	51.45	42.47	26.57	14.74	19.06	31.66	34.61	33.78
2011	AE	49194	50184	49071	43034	28985	17076	8417	9679	10987	15352	21049	18404
	AP	74.89	76.40	74.70	65.51	44.12	26.00	12.81	14.73	16.73	23.37	32.04	28.02
2012	AE	39159	44762	44790	32913	27312	19012	10393	8252	13459	18231	21602	36737
	AP	59.61	68.14	68.19	50.10	41.58	28.94	15.82	12.56	20.49	27.75	32.89	55.93
2013	AE	46725	53109	49531	39807	32037	19240	8560	7636	10346	16924	26874	24235
	AP	71.13	80.85	75.40	60.60	48.77	29.29	13.03	11.62	15.75	25.76	40.91	36.89
2014	AE	43607	49454	44383	39776	30283	20262	11187	7884	15886	16083	25353	32098
	AP	66.38	75.28	67.57	60.55	46.10	30.85	17.03	12.00	24.18	24.48	38.60	48.86
2015	AE	42186	46088	49649	45433	33722	23776	15032	9128	14771	23932	29538	36485
	AP	64.22	70.16	75.58	69.16	51.34	36.19	22.88	13.90	22.49	36.43	44.97	55.54
2016	AE	36381	39952	38001	34882	25132	15518	9387	7240	9512	11047	13699	23645
	AP	55.38	60.82	57.85	53.10	38.26	23.62	14.29	11.02	14.48	16.82	20.85	36.00
2017	AE	38083	54522	45814	41340	33074	23365	13097	9077	11895	12247	20436	34301
	AP	57.97	83.00	69.74	62.93	50.35	35.57	19.94	13.82	18.11	18.64	31.11	52.22
2018	AE	35647	34813	35798	29552	23288	13254	9320	8140	18336	33718	37651	37209
	AP	54.27	53.00	54.50	44.99	35.45	20.18	14.19	12.39	27.91	51.33	57.32	56.64
2019	AE	45084	49393	49220	41178	29611	25537	16881	12354	12071			
	AP	68.63	75.19	74.93	62.69	45.08	38.88	25.70	18.81	18.38			

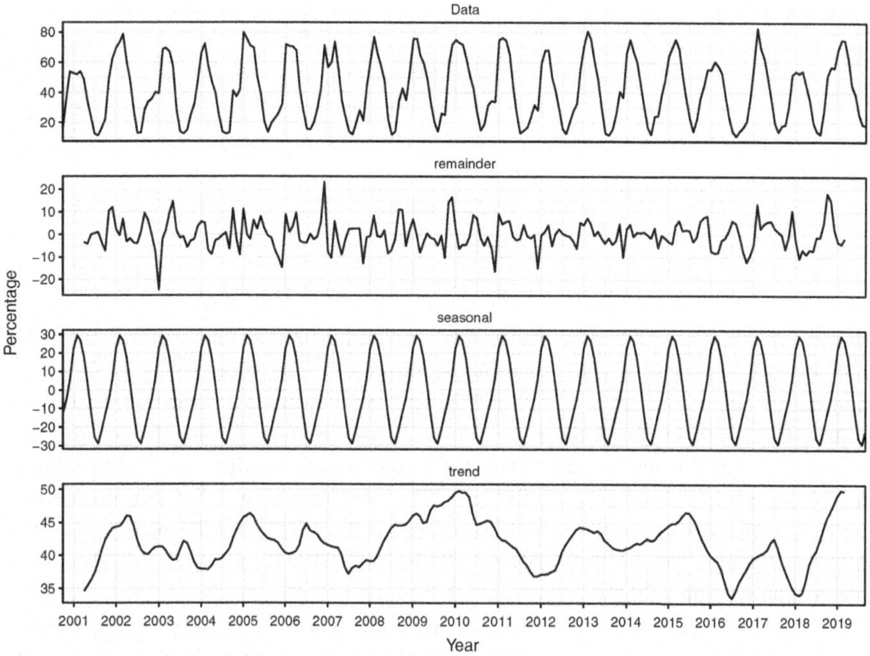

Fig. 3.16 Additive decomposition of the MODIS SCA in the study area of Ladakh over the period of 2000–2019

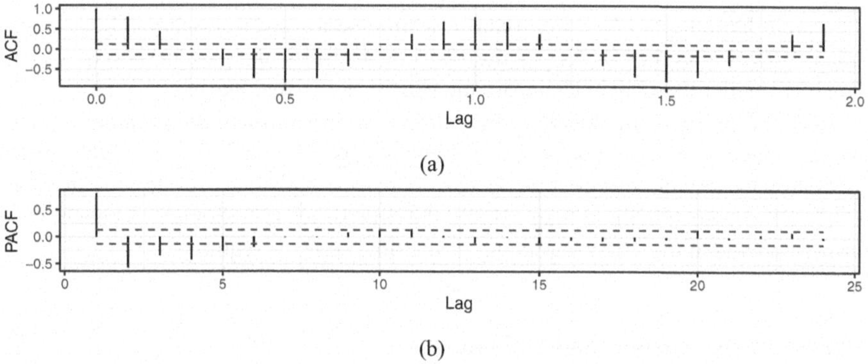

Fig. 3.17 (**a**) ACF and (**b**) PACF correlograms of the MODIS SCA in the study area of Ladakh for the observation period of 2000–2019

To get an idea of the stationarity of the time series initially Autocorrelation Function (ACF) and Partial Autocorrelation Function (PACF) tests were conducted (Box et al. 2008; Chatfield and Xing 2019). Figure 3.17 shows plots of the results.

A 95% confidence interval is shown with a dashed blue line. If the autocorrelation crosses this line, it means that specific lags are significantly correlated with the

Table 3.5 ADF test of the MODIS SCA in the study area of Ladakh from 2000–2019

Test	Test statistic	*p*-value
ADF	−9.2725	0.01

Table 3.6 KPSS test of the MODIS SCA in the study area of Ladakh for the observation period of 2000–2019

Test	Test statistic	*p*-value
KPSS	0.0134	0.1

current series. The ACF plot displays a sinusoidal or periodic pattern, which is evidence of seasonality, and no trend is observed. The time series is therefore stationary. The peak lags at 0, 12, and 24 (months) indicate a positive relationship with a 12 month cycle of the time series. In this regard, 12 are the appropriate period units S for the model. The PACF plot confirms the seasonality with a decreasing periodic pattern over the lags. The obvious spike at lag 1 indicates that the time series follows a first order Autoregressive (AR(1)) model process and the appropriate value for p is 1 (McMurry and Politis 2010; Hyndman 2015; Hyndman and Athanasopoulos 2021).

To further check the stationarity of the time series subsequently Augmented Dickey-Fuller (ADF) and Kwiatkowski-Philips-Schmidt-Shin (KPSS) tests were conducted (Dickey and Fuller 1979; Kwiatkowski et al. 1992; Fuller 1995). Table 3.5 shows the output of the ADF test.

As a rule of thumb, where the p-value < 0.05 (5%), the test is in favour of the alternative hypothesis (H_1), i.e., data in a time series are stationary (Dickey and Fuller 1979; Fuller 1995). As per the above test results, the p-value is 0.01 (1%), which is smaller than 0.05. Therefore the time series is stationary.

Table 3.6 presents the details of the KPSS test. Here, where the p-value is more than 5%, the test is in favour of the null hypothesis (H_0), i.e., data in a time series are stationary (Kwiatkowski et al. 1992). As per the above test results, the p-value is 0.1 (10%), which is more than 5%. So the time series is stationary. As such, it is further concluded that no differencing of the time series is required.

ARIMA Model Analysis and Validation

In this research an Autoregressive Integrated Moving Average (ARIMA) model was chosen (Box et al. 2008; Hyndman and Athanasopoulos 2021), to best fit the time series of the snow cover. Table 3.7 summarises the calculated parameters.

Table 3.7 ARIMA model parameters for the study area of Ladakh from 2000–2019

Model	Parameters
ARIMA	(1,0,0)(2,1,0)[12]

Table 3.8 AIC_c values of alternative SARIMA models, applied to the original time series. The chosen SARIMA (1,0,0)(2,1,0)[12] model has the lowest AIC_c value of 1530.29 (shown in red)

Model	AIC_c	Model	AIC_c
ARIMA(1,0,0)(2,1,0)[12]	1530.29	ARIMA(0,0,1)(1,1,0)[12]	1572.24
ARIMA(0,0,3)(2,1,0)[12]	1531.34	ARIMA(0,0,2)(1,1,0)[12]	1572.72
ARIMA(3,0,0)(2,1,0)[12]	1532.11	ARIMA(0,0,0)(2,1,0)[12]	1583.95
ARIMA(1,0,1)(2,1,0)[12]	1532.19	ARIMA(0,0,3)(0,1,0)[12]	1612.43
ARIMA(1,0,2)(2,1,0)[12]	1532.58	ARIMA(1,0,2)(0,1,0)[12]	1613.35
ARIMA(2,0,1)(2,1,0)[12]	1533.72	ARIMA(1,0,1)(0,1,0)[12]	1613.62
ARIMA(0,0,2)(2,1,0)[12]	1537.72	ARIMA(1,0,3)(0,1,0)[12]	1613.84
ARIMA(0,0,1)(2,1,0)[12]	1538.79	ARIMA(0,0,4)(0,1,0)[12]	1613.96
ARIMA(0,0,3)(1,1,0)[12]	1568.47	ARIMA(3,0,0)(0,1,0)[12]	1614.08
ARIMA(1,0,0)(1,1,0)[12]	1568.85	ARIMA(4,0,0)(0,1,0)[12]	1614.51
ARIMA(2,0,2)(1,1,0)[12]	1569.04	ARIMA(1,0,0)(0,1,0)[12]	1615.03
ARIMA(3,0,1)(1,1,0)[12]	1569.28	ARIMA(2,0,0)(0,1,0)[12]	1615.23
ARIMA(3,0,0)(1,1,0)[12]	1569.34	ARIMA(0,0,5)(0,1,0)[12]	1615.77
ARIMA(1,0,2)(1,1,0)[12]	1569.53	ARIMA(0,0,2)(0,1,0)[12]	1615.79
ARIMA(1,0,3)(1,1,0)[12]	1570.13	ARIMA(0,0,1)(0,1,0)[12]	1616.43
ARIMA(1,0,1)(1,1,0)[12]	1570.41	ARIMA(5,0,0)(0,1,0)[12]	1616.51
ARIMA(0,0,4)(1,1,0)[12]	1570.42	ARIMA(0,0,0)(1,1,0)[12]	1624.62
ARIMA(2,0,0)(1,1,0)[12]	1570.68	ARIMA(0,0,0)(0,1,0)[12]	1688.87
ARIMA(4,0,0)(1,1,0)[12]	1570.99		

A breakup of the model parameters shows that a seasonal component was determined. This was expected due to the seasonal nature of the snowfall. The model is therefore a Seasonal Autoregressive Integrated Moving Average (SARIMA) (Chatfield and Xing 2019; Hyndman and Athanasopoulos 2021). It has parameters 1,0,0 for, respectively, autoregressive aspect p, amount of differencing d and moving average q; 2,1,0 for, respectively, P, D, Q, corresponding to the seasonal components of the model; and lastly 12 (months) for period units S. An indication of the seasonal nature and the values of p and S was already predicted in the analysis of the decomposition and the ACF and PACF plots in the previous section.

To obtain the above model, the $(p, d, q)(P, D, Q)[S]$ parameter combination that returned the lowest Akaike's Information Criterion (AIC) or Corrected Akaike's Information Criterion (AIC_c) values was examined (Akaike 1973, 1974; Cavanaugh 1997; Burnham and Anderson 1998). This validation was carried out by observing the change of other parameters against constant d and D values. AIC_c showed the lowest values, shown in Table 3.8 together with alternative $(p, d, q)(P, D, Q)[S]$ parameter combinations.

The graphical output of the completed tests on the residuals of the SARIMA model is illustrated in Fig. 3.18. The residual plots appear to be centred around zero mean as white noise, with no discernable pattern (uncorrelated). The generally high p-values for the Ljung-Box test also demonstrate that the residuals are similar to

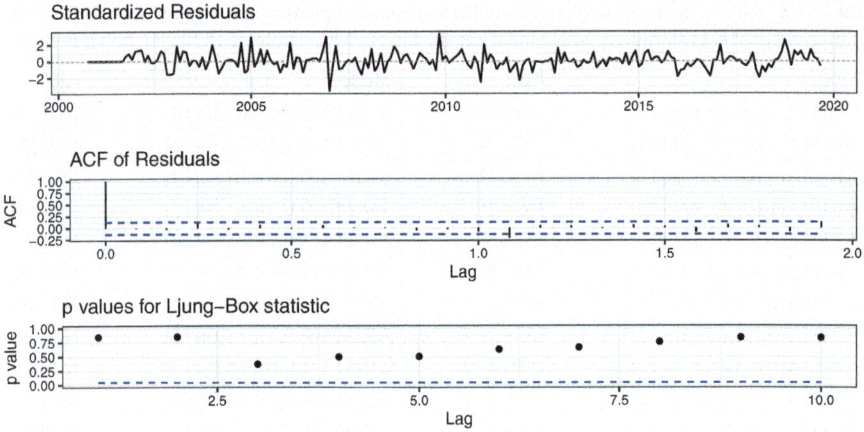

Fig. 3.18 Residual test plots of the SARIMA model. From top to bottom are shown a time series, ACF correlogram and *p*-values plot of the Ljung-Box test (Study area of Ladakh over the period of 2000–2019)

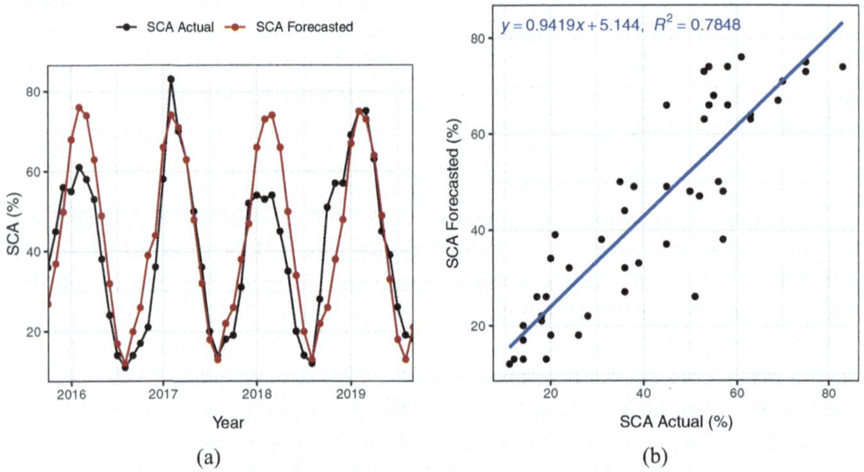

Fig. 3.19 Comparison of the forecasted SCA monthly mean percentages against the actual MODIS SCA. (**a**) Time series model, with the forecasted values in red and actual values in black; (**b**) Linear regression model, with the line of best fit between the forecasted and actual values in blue

white noise. This verified the suitability of the model (Box and Pierce 1970; Ljung and Box 1978; Hyndman and Athanasopoulos 2021).

The forecasted and the actual monthly mean percentages of the MODIS SCA data in the period October 2015–September 2019 were fitted well, as the coefficient of determination (R^2) achieved 0.78 (see Fig. 3.19). This further increased the confidence in the SARIMA model to represent the mean monthly SCA in the study area, and it could be used to forecast the trend of future snowfall.

3.6 Field Survey Measurements

In this section the outcome of snow sampled on a snowboard in Khakshal village (period 2015–2019), in situ snow surveys in South Phulu (period 2015–2018), snow stake camera recordings in South Phulu (winter 2016/2017), and the Gyamtsa (period 2015–2017) and Khakshal automatic weather stations (AWSs) weather variables (period 2016–2019) are presented.

3.6.1 Snowboard Measurements

The snow depth and snow water equivalent (SWE) were measured manually using a snow tube and snow weighing spring balance on the snowboard. Additionally, the depth of the snow was also obtained by recording a graduated ruler with an automated camera in Khakshal. The resulting data are plotted in Fig. 3.20. See also Table B.

The manually sampled snow depth on the snowboard, for three consecutive snow hydrological years in the period between 2015 and 2018, shows a variation in snowfall. The SWE could not be measured for snow depths $\leq$ 1 cm.

During the first snow hydrological year 2015/2016, the snow onset started in November 2015. The results for 12 snowfall days show a (point measurement mean) minimum snow depth of 0.4 cm on 27 November 2015; a maximum snow depth of 4.6 cm and 3 mm SWE on 5 January 2016. The total amount of snowfall was 24.6 cm, with an overall mean of 2.1 cm and the total SWE amounted to 13 mm, with overall mean of 2 mm.

The second snow hydrological year 2016/2017 showed a delayed onset of snowfall, starting only in January 2017, but it was an unprecedented snow year. Not always could the snowfall be measured manually. Results for the measured 10 days of snowfall: A minimum snow depth of 0.2 cm on 2 and 8 January 2017; a maximum of 10 cm snow depth and 13 mm SWE on 6 April 2017. The total amount of snowfall was 27.4 cm, with an overall mean of 2.7 cm and the total SWE amounted to 34 mm, with overall mean of 5 mm.
The Raspberry Pi (RPi) camera (Raspberry Pi Foundation 2016) also started this year to automatically record snow depth measurements of the snowboard and showed a good agreement with the manually sampled snow depth. The results give a complete overview of snowfall events: During the measured 25 days, the same minimum and maximum snow depth values as above, but the total amount of snowfall was 66.3 cm, with an overall mean of 2.7 cm, making it the highest snow depth compared to all the other years.

During the third snow hydrological year 2017/2018, the snow onset started already in November 2017; however, it was a comparatively dry (very little snowfall) and short snow season. Here too, not always was the snow measured manually. Results for the measured 4 days: A minimum snow depth of 0.4 cm on 21 April

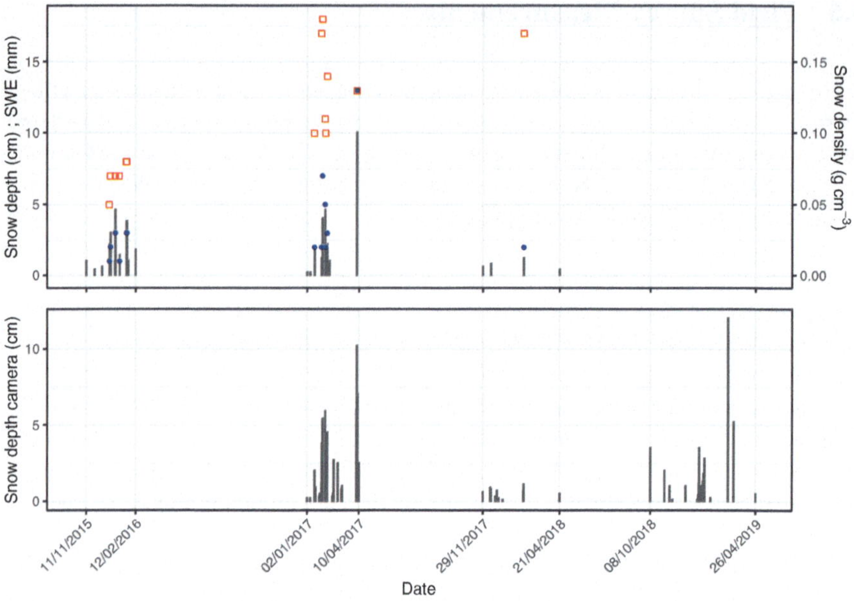

Fig. 3.20 Snowboard data of Khakshal in the period of 2015–2019. The start and end dates of each snow hydrological cycle are shown on the x-axis. Snow depth (in cm) is shown in vertical grey bars, SWE (in mm) in blue dots, and snow density (in g cm^{-3}) in red squares. In the upper chart a snow tube was used to measure the snow depth, and a snow spring balance for the SWE. In the lower chart the snow depth was automatically recorded with a Raspberry Pi (RPi) Camera Module

2018; maximum snow depth of 1.2 cm with 2 mm SWE (the only value that could be calculated) on 13 February 2018. The total amount of snowfall was 3 cm, with an overall mean snow depth of 0.8 cm.

The RPi camera showed for the 10 snowfall days a minimum snow depth of 0.1 cm on 5 January 2018, and a maximum snow depth of 1.1 cm on 13 February 2018. The total amount of snowfall was 5.9 cm, with an overall mean snow depth of 0.6 cm.

Since manual snow sampling was not carried out for the snow hydrological year 2018/2019, all information regarding the snow depth was obtained from the RPi camera. The snow onset started quite early in October 2018, with for the 17 snowfall days a minimum snow depth of 0.1 cm on 5 January 2019 and a maximum of 12 cm on 4 March 2019. The total amount of snowfall was 37.6 cm, with an overall mean snow depth of 2.2 cm.

The snow density showed differences between the first two snow hydrological years, with a minimum of 0.05 g cm^{-3} on 25 December 2015 and 0.10 g cm^{-3} on, respectively, 16 January and 6 February 2017; a maximum of 0.08 g cm^{-3} on both 26 and 27 January 2016, and 0.18 g cm^{-3} on 31 January 2017; and average snow densities of 0.07 and 0.13 g cm^{-3} for, respectively, 2015/2016 and 2016/2017. Only one density value could be calculated for 2017/2018, a value of 0.17 g cm^{-3} on 13 February 2018.

3.6.2 *Snow Survey Observations*

The snow measurements completed at the three different snow plots in South Phulu are plotted in Fig. 3.21 (see also Table C). Plot 1 measurements of snow depth and SWE were carried out throughout the winters of 2015/2016, 2016/17, and 2017/18. There is however no data available for the winter periods of 2016/17 and 2017/18 for plots 2 and 3, due to the hazardous accessibility of their higher terrain, except on 7 January 2017.

Plot 1 (4770 m a.s.l.)

The first series of snow surveys during winter 2015/2016, conducted five times in the months December-February on plot 1, shows an average of 28 cm from 30–35 random point measurements, varying between 11 cm and 45 cm of snow depth with an SD mean of 5.52. The SWE varies from minimum of 17 mm to maximum of 78 mm, with a mean of 44 mm. A mean snow density of 0.16 g cm^{-3}, with minimum and maximum of 0.14 g cm^{-3} and 0.17 g cm^{-3}, respectively, was observed.

The second series of snow surveys during winter 2016/2017 took place later in the months of January, February, and March, due to a delay in snow accumulation. The mean snow depth of random point measurements was 51 cm, varying from 10 cm to 86 cm, with an SD mean of 4.73. A mean SWE of 79 mm, with a minimum of 18 mm and maximum of 128 mm was detected. The mean snow density was 0.16 g cm^{-3}, varying between minimum of 0.14 g cm^{-3} to maximum of 0.17 g cm^{-3}. The

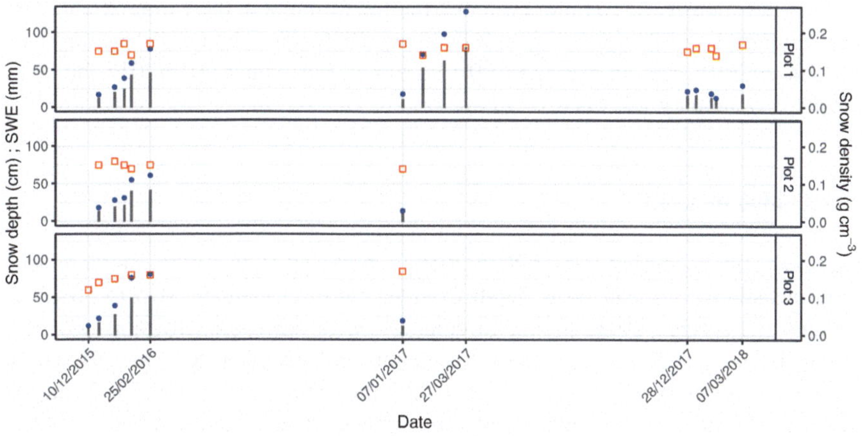

Fig. 3.21 In situ snow survey data of the three snow plots at South Phulu during the observation period of 2015–2018. The start and end dates of the surveys for each snow hydrological cycle are shown on the x-axis. Snow depth (in cm) is shown in vertical grey bars, SWE (in mm) in blue dots, and snow density (in g cm^{-3}) in red squares. Plot 1 is at an elevation of 4770 m a.s.l., plot 2 at 4810 m a.s.l., and plot 3 at 4860 m a.s.l.

rapid increase rate in both mean snow depth and SWE was due to frequent snowfall in the month of January, and snow redistribution by wind.

Snow surveys during winter 2017/18 were conducted from December 2017 to March 2018. Observed were a mean snow depth of 14 cm from 35 random point measurements, varying between 9 cm and 17 cm, with an SD mean of 3.75. An SWE mean of 22 mm with a minimum of 13 mm and maximum of 30 mm. A snow density mean of 0.16 g cm^{-3}, ranging from 0.14 g cm^{-3} to 0.17 g cm^{-3}.

Overall the snow density shows a fairly even distribution throughout the three winters, with a minimum of 0.14 g cm^{-3} in February 2016, 2017 and 2018, and a maximum of 0.17 g cm^{-3} in January 2016, February 2016, 7 January 2017 and 7 March 2018, with an average snow density for each winter of 0.16 g cm^{-3}.

Plot 2 (4810 m a.s.l.)

The results of the snow surveys during winter 2015/2016 on plot 2 show the snow depth of 27 cm of 30–34 of random point measurements, varying between 12 cm and 40 cm, with an SD mean of 3.87. An average SWE of 41 mm, varying between 18 mm to 61 mm was observed, and the density of snow showed a mean of 0.15 g cm^{-3}, with a minimum and maximum of 0.4 g cm^{-3} and 0.16 g cm^{-3}, respectively.

The constant rate or no visible effect on the snow depth and SWE after the peak snowfall in January 2016 was due to the snow cover surface being scoured by wind and redistribution into the gully of the nearby stream.

During winter 2016/2017 point measurements were only conducted on 7 January 2017, with a point measurement mean snow depth of 9.9 cm and an average SWE of 14 mm.

The snow density has a quite even distribution throughout the winter 2015/2016, with a minimum of 0.14 g cm^{-3} on 2 February 2016, maximum of 0.16 g cm^{-3} on 12 January 2016, and an average of 0.15 g cm^{-3}. For winter 2016/2017 the snow density of 7 January 2017 is 0.14 g cm^{-3}.

Plot 3 (4860 m a.s.l.)

The snow surveys during winter 2015/2016 on plot 3 show that it has an increasing point measurement mean snow depth from 10 cm in December 2015 to 50 cm at the end of February 2016, higher in comparison to the other two plots. For this winter the average SWE varies between 12 mm in December 2015 and 81 mm at the end of February 2016, again higher than the other plots. Like plot 2, during winter 2016/2017 measurements were only conducted on 7 January 2017.

The significant amount of snow cover and higher values for the snow depth and SWE of plot 3 are due to its location in a predominantly shaded area, and the snow deposits additionally due to wind redistribution.

The snow density shows slightly increasing throughout the winter 2015/2016, with a minimum of 0.12 g cm^{-3} on 10 December 2015, maximum of 0.16 g cm^{-3}

on 2 and 25 February 2016, and an average of 0.15 g cm^{-3}. For winter 2016/2017 the snow density of 7 January 2017 is 0.17 g cm^{-3}.

To compare the snow surveys conducted during the first winter of 2015/2016 on three different plots varying in altitude, angle, and shadow area zone: The mean SWE remained initially quite similar with little variation (17 mm, 18 mm, and 22 mm, respectively, for plot 1, 2, and 3), and thereafter gradually started to increase over time. However, during the end of this winter, the SWE of plot 2 had increased at a lower rate to a maximum amount of 61 mm in comparison with plots 1 and 3 that measured 78 mm and 81 mm, respectively. The lower SWE of plot 2 was due to the redistribution of snow by wind, which was also witnessed during the field surveys. Furthermore, on the same dates the snow density at the three snow survey plots remained quite consistent, with an average of 0.16 g cm^{-3} and varying between a minimum of 0.14 g cm^{-3} and a maximum of 0.17 g cm^{-3}.

3.6.3 South Phulu Snowpack

The South Phulu camera recorded photographs of a graduated snow stake to interpret the spatio-temporal patterns of snow cover, snow depth and snowfall interception in the near range of the camera and monitored the Phutse glacier in the far range of the South Phulu watershed area. Figures 3.22 and 3.23 are examples of hourly and mid-monthly photographs of the graduated snow stake positioned at the South Phulu location during the winter 2016/2017. Clearly the accumulation and ablation of the snowpack can be seen over the months, in an area that would remain inaccessible in the period from January to May.

The snowpack in Fig. 3.24 accumulated from 0 cm snow depth on 1 January 2017, to a maximum snow depth of 79 cm on 10 April, to gradually melt again to 0 cm snow depth on 26 May, with an overall mean of 32 cm. The mean trend line (in blue) gives a good idea of the average snow depth during the measured period. For example, due to a lack of snowfall the snow packed somewhat in the

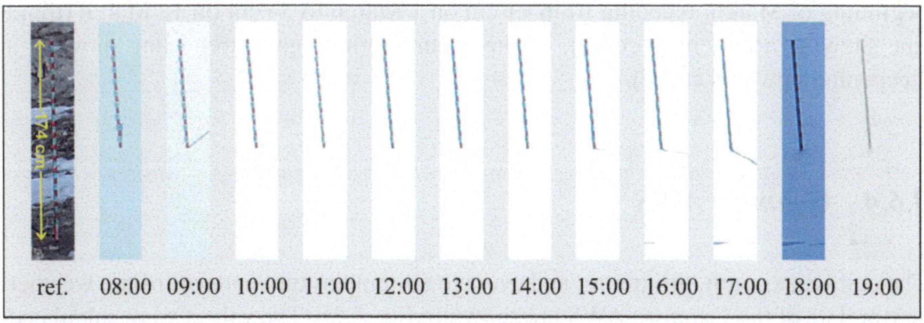

Fig. 3.22 Hourly photos of the snow stake recorded by the South Phulu camera on 10 April 2017 (A photo taken in December 2016 is used for reference)

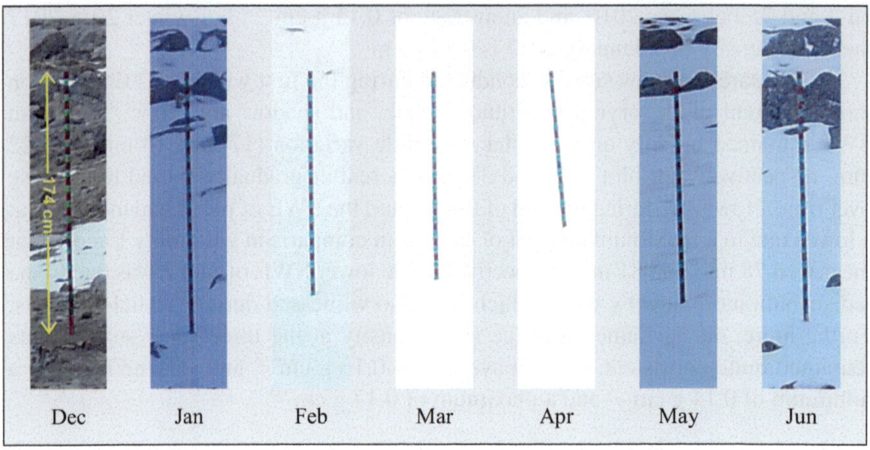

Fig. 3.23 Mid-monthly photos of the snow stake recorded at noon by the South Phulu camera during winter 2016/2017

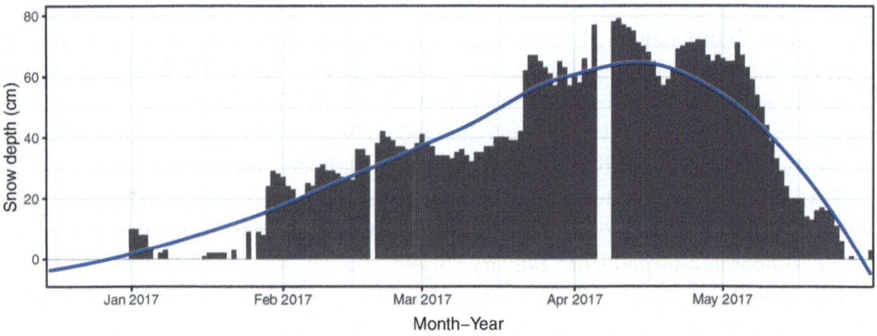

Fig. 3.24 Daily maximum snow depth (in cm), shown with vertical grey bars, recorded by the South Phulu camera in the period from December 2016 until May 2017. The mean trend line is shown in blue

beginning of March, receding from 41 cm on 1 March to 34 cm on 12 March (trend line showing a flattening curve), before accumulating again (trend line showing a steepening upwards curve).

3.6.4 Gyamtsa AWS

The daily maximum, minimum, and average air temperature patterns of the weather sensor data of the Gyamtsa AWS are shown in Fig. 3.25. The values were calculated at an interval of 1 hour, in the period from 1 December 2015 until 21 February 2017, at an elevation of 4220 m a.s.l.

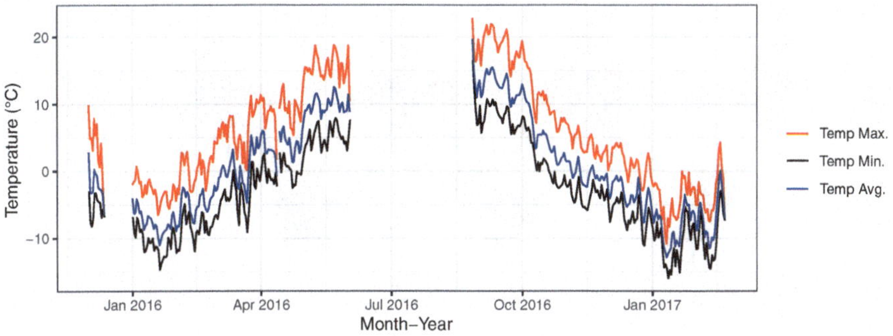

Fig. 3.25 Gyamtsa AWS daily maximum, minimum & average air temperature data, in the observation period 1 December 2015 until 21 February 2017

The data shows some gaps due to limitations in the solar panel set-up, software coding, heavy snowfall events, and prolonged cloudy weather. Due to these gaps and the short recording period in general, it was not possible to analyse seasonal trends or variations. However, some of the data were used in a correlation analysis with the monthly mean RSLE (see Sect. 3.7).

To summarise the measured period, the daily air temperature varied between as high as a 22.8 °C maximum in summer 2016 and as low as a −16.1 °C minimum in winter 2016/2017. The highest and lowest daily average temperatures were 19.7 °C, measured on 27 August 2016, and −12.9 °C, measured on 11 January 2017.

3.6.5 Khakshal AWS

Figure 3.26 illustrates the daily maximum, minimum and average air and soil temperature patterns of the weather sensor data of the Khakshal AWS in detail. The values were calculated at an interval of 5 minutes, in the period from 27 May 2016 until 30 September 2019, at an elevation of 3660 m a.s.l.

The data showed some gaps in 2016 and 2017, due to limitations in the required internet connection for the station and electrical power supply, and occasional breaks due to modifications and maintenance. Some of the data was used in a correlation analysis with the monthly mean RSLE (see Sect. 3.7).

The daily air temperature varied between as high as a 30 °C maximum in summer 2018, to as low as a −19.7 °C minimum in winter 2018/2019. For hydrological year 2016–2017 the seasonal monthly trend showed a highest maximum of 29 °C in August and lowest minimum of −18.9 °C in January. Due to the data gaps in 2016 and 2017, no yearly average air temperature value could be calculated here. For hydrological year 2017–2018, a highest maximum of 30 °C in July, lowest minimum of −18.2 °C in January and annual average of 5.8 °C was observed. Hydrological year 2018–2019 showed a highest maximum of 27.6 °C in July, lowest minimum of

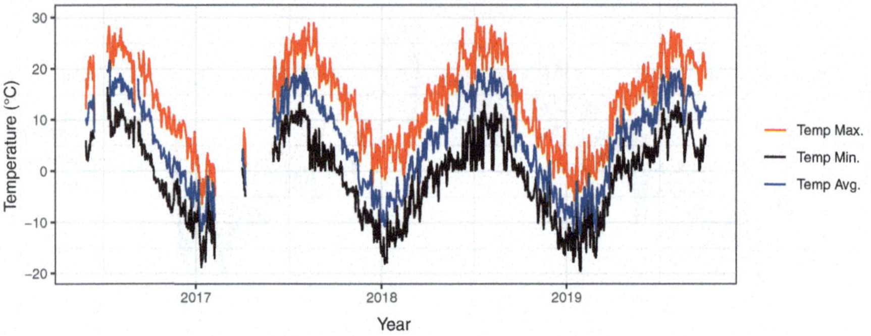

Fig. 3.26 Khakshal AWS daily maximum, minimum & average air temperature data, over the period 27 May 2016 until 30 September 2019

−19.7 °C in January and has an annual average of 4.3 °C. The highest and lowest daily average temperatures were 21.6 °C, measured on 14 July 2016, and −12.5 °C, measured on 28 January 2019.

The daily soil temperature (at 10 cm below the surface) showed a variation of a highest 38.5 °C maximum in summer 2017, and a lowest −14.8 °C minimum in winter 2016/2017. For hydrological year 2016–2017 the seasonal monthly trend showed a highest maximum of 38.5 °C in July and lowest minimum of −14.8 °C in January. Due to the data gaps in 2016 and 2017, no yearly average soil temperature value could be calculated here. For hydrological year 2017–2018, a highest maximum of 31.8 °C in July, lowest minimum of −10.8 °C in January and annual average of 9.8 °C was observed. Hydrological year 2018–2019 showed a highest maximum of 28.8 °C in August, lowest minimum of −11.8 °C in January and has an annual average of 7.8 °C. The highest and lowest daily average soil temperatures were 26.3 °C, measured on 11 July 2018, and −8.7 °C, measured on 13 January 2017.

Collected at an elevation of 3665 m a.s.l. in the period from 27 May 2016 until 30 September 2019, the wind data of the Khakshal AWS are plotted in wind rose charts. Figure 3.27a illustrates the yearly wind speed and direction, in a range of 0–93.7 km hr^{-1}, distributed in five wind speed categories of different colour. The maximum wind speed of 93.7 km hr^{-1} was recorded during a wind gust on 17 November 2017, and the data has an overall mean wind speed of 4.2 km hr^{-1}.

The wind rose pattern of 2016 reveals a dominant percentage between 30% and 40% of the total wind blowing from the north. Wind from the north-east, east, south-east, south, south-west, west, and north-west shows approximately equal to or less than 10%. Overall wind blowing from a western direction is very minimal.

The wind roses of the years 2017, 2018, and 2019 show similar characteristics, with the wind blowing predominantly from the north, at around 30% to 40% of the total. However, for all these years the percentage of wind blowing from the north-west shows a little more than 10%, and the concentration of the highest wind speed

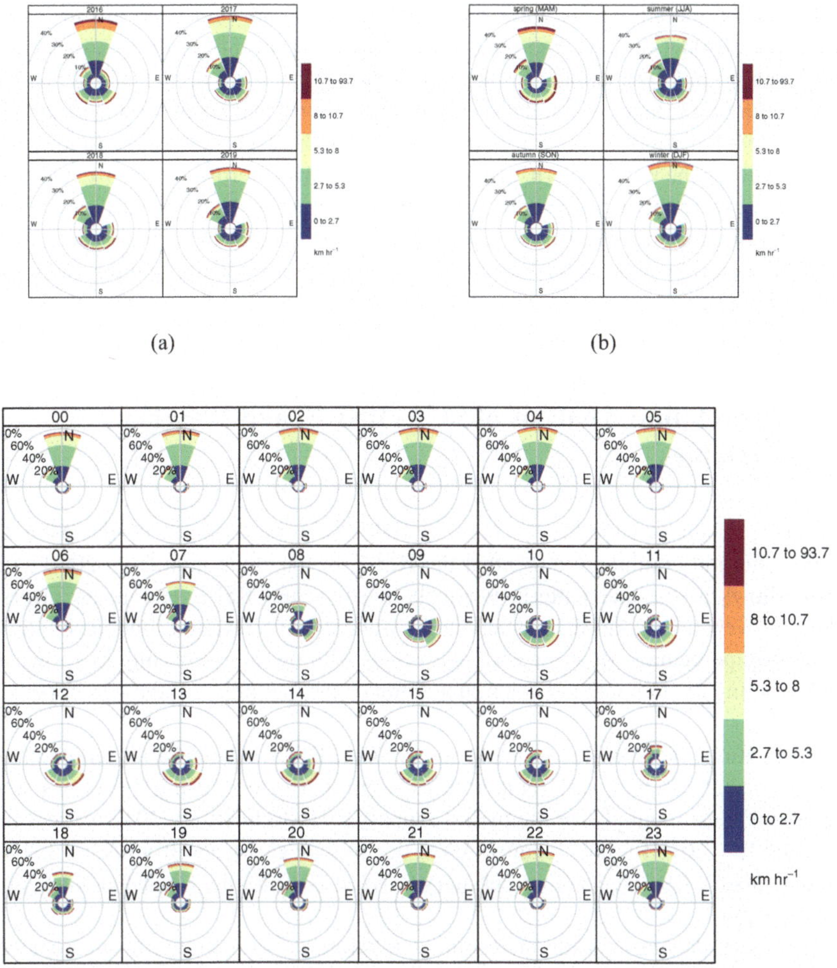

(c)

Fig. 3.27 Weather sensor data of the Khakshal AWS in the period 27 May 2016–30 September 2019, showing wind roses of (**a**) yearly, (**b**) seasonal, and (**c**) hourly wind data

category (10.7 to 93.7 km hr^{-1}) is very negligible or not noticeable at all compared to 2016.

Figure 3.27b shows the seasonal wind speed and direction of the periods spring (March–May), summer (June–August), autumn (September–November), and winter (December–February). The intensity of high wind speeds in the spring season is very high. The wind directions however show a similar pattern to the yearly wind rose plot, except in the summer season, when the percentages of wind blowing from the north show a decreased range of 20–30%.

The hourly details of the wind speed and direction are presented in Fig. 3.27c. In the period 00–06 hours, the wind blows mostly from the north, at around 40–60%, and around 20% from a north-western direction. In the period 07–08 hours a decrease in percentages of wind blowing from the north and north-west is observed, and a little bit of wind blowing from the east and south-east is noticeable. In the early morning hours the wind pattern changes drastically from north-east to east, and then to south-west. In the daytime wind blows from a direction between east and south-west, dominating mostly towards a southern rather than northern direction. Lastly, the wind direction changes back to the north and north-west in the evening and night (18–23 hours).

3.7 Correlation Analysis of SCA, RSLE, and Weather Variables

The comparison of monthly average MODIS SCA percentages of the MODIS spatio-temporal combination, against the RSLE in the study area from 2000 until 2019, shows a seasonal pattern, where both snow and RSLE feature an inverse relation over time. An increasing snow transitional elevation line leads to a decreasing percentage of snow and vice versa, with the lines crossing in spring or the beginning of summer, and again in autumn or the beginning of the winter period (Fig. 3.28a).

Correlations between the monthly mean SCA, RSLE, Gyamtsa air temperature and Khakshal air and soil temperatures were performed using Pearson's and Spearman's test methods for the period January 2016 to February 2017 (except for June and July 2016 for the Gyamtsa AWS, due to data gaps), and from August 2016 to September 2019 (except from February to April 2017 for the Khakshal AWS, due to data gaps).

The SCA decreases with increasing temperature, with a significant negative correlation ($p < 0.05$ for Gyamtsa and $p < 0.01$ for Khakshal). Pearson's correlation test for the SCA against air temperature shows values of –0.69 for Gyamtsa and –0.79 for Khakshal, and –0.75 for SCA against the Khakshal soil temperature. Spearman's correlation test for the same comparisons shows slightly higher values of, respectively, –0.72, –0.80, and –0.78. See Fig. 3.28b (upper figure) and Table 3.9.

The RSLE shifts upward with increasing temperature, with a significant positive correlation ($p < 0.01$). Pearson's correlation test for the RSLE against air temperature shows values of 0.84 for Gyamtsa and 0.89 for Khakshal, and 0.85 for the RSLE against the Khakshal soil temperature. Spearman's correlation test for the same comparisons shows the similar values 0.85, 0.88, and 0.85 (Fig. 3.28b (lower figure) and Table 3.9).

Both tests demonstrate that the RSLE is marginally more correlated to temperature than the SCA is.

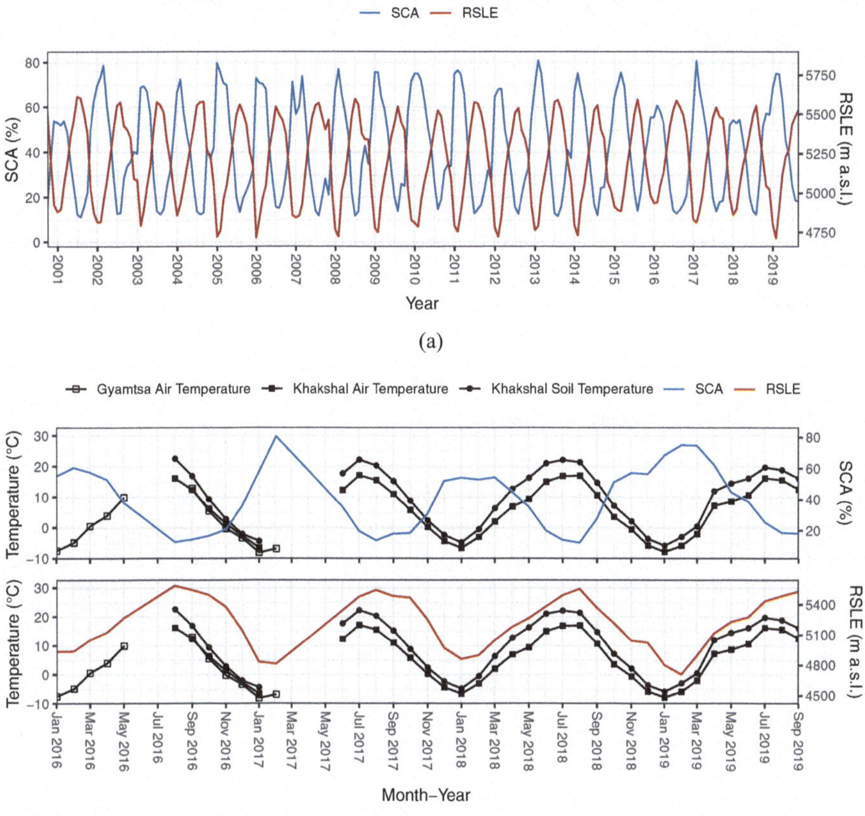

Fig. 3.28 (**a**) MODIS monthly mean SCA versus RSLE (period 2000–2019); and (**b**) MODIS monthly mean SCA (upper figure) and RSLE (lower figure) versus Gyamtsa AWS (Jan 2016–May 2016 and Sep 2016–Feb 2017) and Khakshal AWS (Aug 2016–Jan 2017 and Jun 2017–Sep 2019) monthly mean temperature variables

Table 3.9 Correlation between monthly average SCA, RSLE and Gyamtsa temperature (GT), Khakshal temperature (KT) and Khakshal soil temperature (KST)

	SCA vs GT	SCA vs KT	SCA vs KST	RSLE vs GT	RSLE vs KT	RSLE vs KST
Pearson's r	−0.69*	−0.79	−0.75	0.84	0.89	0.85
Spearman's ρ	−0.72*	−0.80	−0.78	0.85	0.88	0.85

Significance level: $p < 0.01$, $^{*}p < 0.05$

3.8 Snow Lines from Gyamtsa and Lungmar Camera Stations

The snow polylines (in red) of cloudless Landsat NDSI snow cover images overlaid on the orthorectified photographs of the south-facing Gyamtsa catchment area in Fig. 3.29 by the Gyamtsa camera station, and the north-facing Stok catchment area in Fig. 3.30 by the Lungmar camera station, were used to delineate snow lines. The lines show a good congruence with the actual snow extent on four different dates (21 January, 6 February, 9 March and 28 May 2016), displayed in the miniature photographs.

3.8.1 Gyamtsa Camera Station

The main photograph taken at the Gyamtsa location (Fig. 3.29a), on an almost snowless day for clarity purposes in this example, shows how the resulting snow lines (in four different colours) look together. 6 February 2016 (in a yellow line with mean elevation 4386 m a.s.l.) shows a snow line of the snow cover with a low elevation in the winter season, and 28 May (in red, mean elevation: 5410 m a.s.l.) shows a snow line with a high elevation, with most of the snow cover melted. 21 January (in blue, mean elevation: 4619 m a.s.l.) features an advancing snow line and 9 March (in green, mean elevation: 4812 m a.s.l.) a retreating snow line.

3.8.2 Lungmar Camera Station

The main photograph taken at the Lungmar location (Fig. 3.30a) illustrates again the resulting snow lines. 21 January (in a blue line with average elevation: 4743 m a.s.l.), 6 February (in yellow, mean elevation: 4240 m a.s.l.) and 9 March 2016 (in green, mean elevation: 4728 m a.s.l.) feature overlapping snow lines of the snow cover with a low elevation in the winter season, and 28 May (in red, mean elevation: 5277 m a.s.l.) shows a snow line with a high elevation in the spring season, with a lot of the snow cover already melted.

A comparison between the south-facing (at Gyamtsa) and north-facing (at Lungmar) snow lines reveals that the south-facing snow lines remain at a higher elevation for all the shown dates in 2016, i.e., on 6 February, 28 May, and 9 March, except 21 January, where the south-facing lines are at a notably lower elevation.

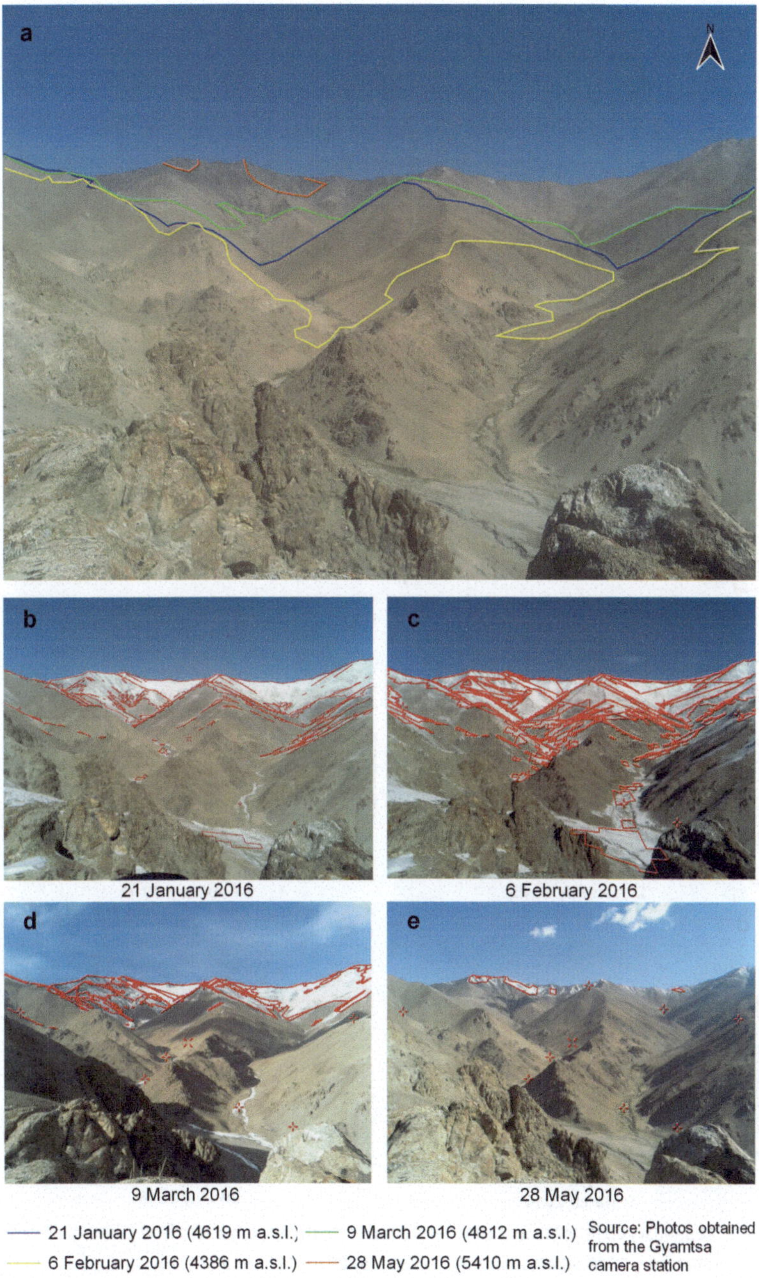

Fig. 3.29 Orthorectified main photograph taken from Gyamtsa on 17 September 2016, with an overview of the resulting snow lines (**a**); Orthorectified photographs and overlaid with snow polylines (red lines) derived from Landsat 8 OLI images of the same dates: 21 January (**b**), 6 February (**c**), 9 March (**d**) and 28 May (**e**) 2016 (Some photos show control points with plus and multiply signs in red)

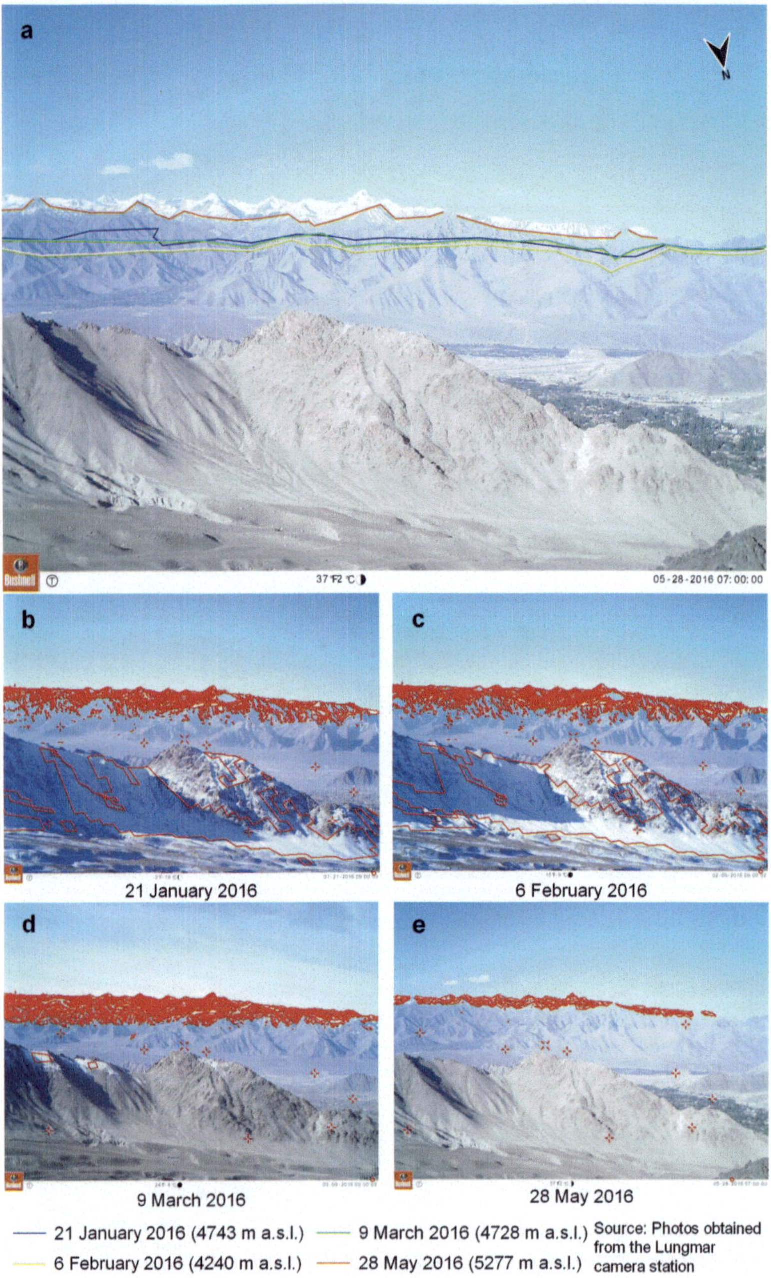

Fig. 3.30 Orthorectified main photograph taken from Lungmar on 28 May 2016, with an overview of the resulting snow lines (**a**); Orthorectified photographs and overlaid with snow polylines (red lines) derived from Landsat 8 OLI images of the same dates: 21 January (**b**), 6 February (**c**), 9 March (**d**) and 28 May (**e**) 2016 (Some photos show control points with plus and multiply signs in red)

References

Akaike H (1973) Information theory and an extension of the maximum likelihood principle. In: Proceeding of the Second International Symposium on Information Theory, Akadémiai Kiadó, Budapest, Hungary, pp 267–281

Akaike H (1974) A new look at the statistical model identification. IEEE Trans Autom Control 19(6):716–723. https://doi.org/10.1109/TAC.1974.1100705

Box GE, Pierce DA (1970) Distribution of residual autocorrelations in autoregressive-integrated moving average time series models. J Am Stat Assoc 65(332):1509–1526. https://doi.org/10.1080/01621459.1970.10481180

Box GE, Jenkins GM, Reinsel GC (2008) Time series analysis: forecasting and control. John Wiley & Sons, Hoboken. https://doi.org/10.1002/9781118619193

Burnham KP, Anderson DR (1998) Practical use of the information-theoretic approach. In: Model Selection and Inference: A Practical Information-Theoretic Approach, Springer, New York, pp 75–117. https://doi.org/10.1007/978-1-4757-2917-7_3

Cavanaugh JE (1997) Unifying the derivations for the Akaike and corrected Akaike information criteria. Stat Probab Lett 33(2):201–208. https://doi.org/10.1016/S0167-7152(96)00128-9

Chatfield C, Xing H (2019) The analysis of time series: an introduction with R. CRC Press, Boca Raton. https://doi.org/10.1201/9781351259446

Cohen J (1960) A coefficient of agreement for nominal scales. Educ Psychol Measur 20(1):37–46. https://doi.org/10.1177/001316446002000104

Cryer JD, Chan KS (2008) Time series analysis: with applications in R, vol 2. Springer, New York. https://doi.org/10.1007/978-0-387-75959-3

Dickey DA, Fuller WA (1979) Distribution of the estimators for autoregressive time series with a unit root. J Am Stat Assoc 74(366a):427–431. https://doi.org/10.2307/2286348

Everitt B, Skrondal A (2010) The cambridge dictionary of statistics, vol 4. Cambridge University Press, Cambridge

Fuller WA (1995) Introduction to statistical time series. John Wiley & Sons, Hoboken. https://doi.org/10.1002/9780470316917

Hall DK, Riggs GA (2007) Accuracy assessment of the MODIS snow products. Hydrol Process 21(12):1534–1547. https://doi.org/10.1002/hyp.6715

Hyndman RJ (2015) Discussion of "High-dimensional autocovariance matrices and optimal linear prediction". Electron J Stat 9(1):792–796. https://doi.org/10.1214/14-EJS953

Hyndman RJ, Athanasopoulos G (2021) Forecasting: principles and practice. OTexts, Melbourne. https://otexts.com/fpp3

Kwiatkowski D, Phillips PC, Schmidt P, Shin Y (1992) Testing the null hypothesis of stationarity against the alternative of a unit root: how sure are we that economic time series have a unit root? J Econom 54(1–3):159–178. https://doi.org/10.1016/0304-4076(92)90104-Y

Ljung GM, Box GE (1978) On a measure of lack of fit in time series models. Biometrika 65(2):297–303. https://doi.org/10.1093/biomet/65.2.297

McMurry TL, Politis DN (2010) Banded and tapered estimates for autocovariance matrices and the linear process bootstrap. J Time Series Anal 31(6):471–482. https://doi.org/10.1111/j.1467-9892.2010.00679.x

Raspberry Pi Foundation (2016) Single-board computers. https://www.raspberrypi.org. Acessed 1 July 2016

Riggs GA, Hall DK, Salomonson VV (2006) MODIS snow products user guide to collection 5. NASA/Goddard Space Flight Center, Greenbelt, Maryland, USA. https://modis-snow-ice.gsfc.nasa.gov/uploads/sug_c5.pdf

Chapter 4
Comparative Perspectives on the Seasonal Snow Patterns in Ladakh

Abstract This chapter presents a comparative discussion of the temporal and spatial distribution of the seasonal snow cover of Ladakh. The monitoring of snow was performed, including ground truthing, using a multi-disciplinary approach, applying the power of advanced remote sensing techniques for the period 2000–2019. Ladakh marks the transitional area between predominantly receding glaciers of the Central Himalaya and some advancing glaciers of the Karakoram, in comparison with the adjoining South Asian Mountain Belt, and it features monsoonal and cold-arid regional climatic differences. Thus, the snow cover is essential to understand the land-atmosphere interaction, and it serves as a primary vital water resource to ensure the availability of water for the irrigation of agricultural fields and other applications.

Keywords Snow · MODIS · NDSI · Cloud reduction · Topographical impact · Forecast · Statistical analysis · Socio-hydrology · Ladakh · India

4.1 MODIS NDSI Threshold Value Calibration and Cloud Removal Approach

Snow cover in the study region of Ladakh consists of often shallow, fragmented snow traces and is ephemeral with complex terrain, making it more difficult to detect than in other regions of the world (Bhutiyani 1999; Hall and Riggs 2007; Gao et al. 2010; Khan and Adams 2019). Consequently, the Moderate Resolution Imaging Spectroradiometer (MODIS) cannot map small snow patches in the transition zone, and this is similar in, e.g., the Tibetan Plateau (Xu et al. 2008; Zhang et al. 2017a) and the Pyrenees (Gascoin et al. 2015). Locally an optimal Normalised Difference Snow Index (NDSI) threshold value for mapping using MODIS snow products is not established so far in the region. Therefore, an accuracy assessment using the highest pixel congruence of Cohen's kappa coefficient (Cohen 1960) and validating the overall MODIS snow data using the Landsat 8 Operational Land Imager (LC8 OLI) relative ground truth method was implemented, to determine a locally optimal NDSI threshold value with the highest overall accuracy (Hall et al. 1995, 2001,

S. Passang, *Snow Cover Distribution and Dynamics*, Advances in Asian Human-Environmental Research, https://doi.org/10.1007/978-3-031-57692-8_4

2002; Salomonson and Appel 2004; Gurung et al. 2011; Thenkabail 2015; Wang 2017; Zhang et al. 2017b; Tran et al. 2019).

A composite methodology of combined Terra, Aqua, and backward–forward temporal and regional snow and land lines incorporated with topographical information was utilised (Parajka and Blöschl 2008; Gafurov and Bárdossy 2009; Parajka et al. 2010; Krajčí et al. 2014; Dong and Menzel 2016). These filters were applied to mitigate the cloud obscuration in MODIS snow cover imagery, with the best matching result as the highest degree of errors were likely to be in the snow transition area consisting of mixed pixels between snow and land, due to the presence of small shallow fragmented snow patches between the accumulation and ablation zones (Parajka et al. 2010).

A number of studies, e.g., Sood et al. (2020), Dharpure et al. (2021), and Muhammad and Thapa (2021), used in the Himalayas the global MODIS NDSI threshold value of 0.40 directly for cloud filtration, in order to detect snow cover pixels, without adjusting to a locally optimal NDSI threshold value, whereas in other studies a much lower threshold of 0.01 or 0.2 was suggested, as the NDSI is highly sensitive towards vegetation, especially the regions with forest cover (Zhang et al. 2010, 2019). In this regard, a combination of the Normalised Difference Vegetation Index (NDVI) and NDSI is used for snow classification in a densely forested area (Thapa et al. 2020). Moreover, Muhammad and Thapa (2021) combined daily MODIS snow cover data with the Randolph Glacier Inventory (RGI) 6.0, which has lots of missing glaciers in the north-eastern and southern parts of the Ladakh region.

Instead of using snow and land transitional elevation Snow Line Mapping approach (SNOWL) filters, some studies (Krajčí et al. 2014; Hauksson 2017; Tran et al. 2019) adopted the snow minimum pixels and meteorological interpolation method proposed by Dong and Menzel (2016) in south-west Germany, which proved very effective in regions with dense snow station networks and regions where the cloud cover remains large and the SNOWL approach cannot be applied. In the case of study region Ladakh however, the initial cloud filter SNOWL method proposed by Parajka et al. (2010) is most suitable, as the region has a very slight cloud cover (out of 365 days about 275 days have clear/almost clear skies), and a shallow, patchy snow cover in the transition zones between snow and land, which was assigned as mixed pixels to reduce uncertainties and enhance the accuracy of the methodology in the region.

According to the experimental outcomes in this study, the optimal NDSI threshold value for the region of Ladakh can be averaged around 0.40, with an overall accuracy rate of around 87.20% for the final MODIS snow data, achieved using the accuracy assessment method of the highest kappa coefficient value. These accuracies match other studies (Klein and Barnett 2003; Hall and Riggs 2007; Liang et al. 2008; Parajka and Blöschl 2008; Gafurov and Bárdossy 2009; Tran et al. 2019; Saydi and Ding 2020), which show accuracy results in the range from 80% to more than 90%. The comparison of MODIS against Landsat data shows underestimation at shallow and ephemeral fragmented snow traces (Hall and Riggs 2007; Gao et al. 2010) and overestimation at permanent snow cover areas, similar to findings of other studies (Paudel and Andersen 2011; Dharpure et al. 2021; Muhammad and Thapa 2021).

4.2 Snow Cover Area and Snow Cover Days

The snow accumulation starts from September and reaches its maximum in February for most subregions of Ladakh, except for Suru and Zangskar where it delays until March. This is in accordance with the findings in another study in Zangskar Valley, where a maximum snow cover area (SCA) is observed in the month of March and the least in August (Khali et al. 2022). However, accumulation begins almost 1 month later in the Karakoram, examined by Dharpure et al. (2021). While the maximum SCA in the Astore basin located to the east of Nanga Parbat was attained in January (Tahir et al. 2015) and in February for the Jhelum basin, located in the Hindu Kush Himalaya (Gurung et al. 2017). Similar to the Suru and Zangskar subregions of Ladakh, the maximum SCA is not attained until March in Gilgit-Baltistan (Bilal et al. 2019) and in the Chenab basin located in Himachal Pradesh (Kour et al. 2016). Ladakh features an SCA annual minimum of 15% in August, maximum of 71% in February, and mean of 42%. Whereas the Chenab basin has an SCA minimum of 60% in August, maximum of 99% in February, and mean of 42%, the same as Ladakh (Jain et al. 2009). In the Astore basin the SCA varies between 7% and 95% (Tahir et al. 2015).

The pixel-wise average snow cover days (SCD) of the Chandra basin was 122 days over the 17-year observation period examined by Dharpure et al. (2020). The mean SCD calculated for the entire Ladakh region was 155 days, and the heterogeneous pattern across seven delineated subregions displayed a maximum of 235 days in Siachan, owed to one of the longest and highest perennial glacier complexes in the world outside the polar regions (Bhutiyani 1999), and a minimum of 87 days in Indus. Suru and Zangskar located in the south-western parts of Ladakh has higher SCD with the average SCD of 184 and 161 in comparison to south-eastern and eastern parts in the subregions of Tshomoriri and Pangong, with averages of 126 and 105 days, respectively. This pattern indicates the decreasing amount of SCA from west to east due to the influence of mid-latitude westerlies and the topography (Ajai et al. 2010).

A significant SCD increasing pixel trend is observed in summer for the subregions of Suru and Zangskar with p-value < 0.1, whereas a significant decreasing trend in the eastern part of Ladakh in the subregions of Pangong, Shayok, Indus, and Tshomoriri can be detected during the winter. The overall annual and summer trends of the snow cover in the entire study area show positive but non-significant, similar to the reports of the Chandra basin (Dharpure et al. 2020).

4.3 Impact of Topography on the SCA and Trends in Seven Subregions

Dharpure et al. (2020) noted that elevation zones between 4000 and 5000 m a.s.l. and above, where glaciers exist, showed a decreasing SCA trend in the Chandra basin. However, the relatively small-sized glaciers in Ladakh are located at altitudes above 5200 m a.s.l. and are regarded as good indicators of climate change (Schmidt and Nüsser 2012). The SCA in Ladakh shows a mostly significant increase in trend for the elevation below 4500 m a.s.l. and a significant decrease in trend for the elevation above 5500 m a.s.l. Moreover, the SCA highly varies between 4500 and 5500 m a.s.l. in all subregions. Thus, the decrease in SCA trends at a higher elevation could have a major impact on water availability.

The study shows that the snow cover has a heterogeneous distribution in all subregions, more or less following the area proportion of the topography, similar to the study in the Western Himalayas by Sood et al. (2020). The dominant SCA in the study region occurred on the north-western- and north-eastern-facing slopes, in comparison with less snow cover on south- and south-western-facing slopes. As north-facing slopes receive less solar radiation, the snow remains in favourable topographical locations on shadowed slopes during autumn and winter (Jain et al. 2009). The south-west-facing slopes are less snow covered than all other slopes, except in the subregion of Suru, where snow cover is slightly less towards the south-facing slopes, whereas the maximum snow cover showed highest on north-facing slopes for Siachan, Indus, and Tshomoriri, east-facing slopes for Shayok, and north-west-facing slopes for Zangskar. This aspect pattern in relation to the SCA was also observed by Khali et al. (2022) in the Zangskar Valley, where the maximum SCA was situated in a north-western orientation followed by north-east-facing slopes. In another study in the Chenab river basin, a higher SCA was detected on the north, north-east, and north-west slopes and less on south and south-western slopes during all seasons, as the higher region's aspect was not a strong regulating factor for snow persistence during the winter. However, it became more significant during the snowmelt (Dharpure et al. 2020).

The slope in relation to SCA across the seven subregions of Ladakh experienced a maximum SCA on the slope zones of 20°–30° for the subregions of Indus and Zangskar, similar to the report of Zangskar Valley investigated by Khali et al. (2022), slope zone of 10°–20° for Siachan and Shayok, and above 40° for the subregion of Suru. A maximum SCA that occurred on the slope zone above 40° was also experienced in the Chenab basin, just adjacent to the subregion of Suru in Ladakh (Dharpure et al. 2020).

4.4 Snowfall Outlook Based on an ARIMA Model

To be able to forecast the trend of future snowfall in the area, a statistical seasonal Autoregressive Integrated Moving Average (ARIMA) model was calculated and validated for accuracy (Box et al. 2008; Hyndman and Athanasopoulos 2021). Test values of a forecast of 4 years (period October 2015–September 2019), tested against the actual data, showed a good agreement, with a coefficient of determination (R^2) of 0.78 in a linear regression analysis.

Comparative studies that describe ARIMA models for the forecasting of snow in the Western Himalayan and Karakoram regions are still very few, with most models being used in rainfall, rainfall-runoff, river, and climate change-/disaster-related studies only (Hassan and Ansari 2010; Khan and Hassan 2015; Mukherjee et al. 2016; Parvaze et al. 2021; Fayaz et al. 2022; Rawat et al. 2022). Notwithstanding, in order to develop a forecasting model that computed the seasonal snow accumulation and depletion in the upper Beas river catchment (north-west Himalaya region), Kumar et al. (2022, 2023) proposed an ARIMA model being the most sufficient to predict future snow. From MODIS MOD10A2 (Terra) and MYD10A2 (Aqua) 8-day composite snow cover products, two appropriate models for both snow accumulation and depletion were prepared and attested, obtaining R^2 values of, respectively, 0.83 and 0.89, in a forecast test period of 2 years (2016–2018). Likewise, Singh et al. (2016) were able to fill in missing snow depth data for several meteorological stations in the Western Himalayas, by making forecasts with an ARIMA model. This enabled the completion of continuous time series for further snow analysis in the study.

The ARIMA model in this research shows good potential to represent future snowfall in the region of Ladakh. Similar workflows of time series analysis and forecasting models could also be implemented for other weather parameters like rainfall, temperature, barometric pressure, relative humidity, wind, etc., crucial for, e.g., future decision-making in hydrological and development policies. The extent of this was beyond the scope of this book, and it is anticipated in the future more investigation will be done in this relatively unexplored field in Ladakh.

4.5 Field Survey and Outdoor Measurements

Snow surveys were conducted at three different plots once a fortnight in South Phulu during the winters of 2015–2016, 2016–2017, and 2018–2019. The snow cover at plot 1, situated at an elevation of 4770 m a.s.l., started to accumulate by the end of December and reached its maximum by the end of January during the first winter of 2015–2016, whereas in the winter of 2016–2017 it reached its peak only in the first week of April. In another study in South Phulu, at an elevation of 4727 m a.s.l., Wani et al. (2021) witnessed a similar snow build-up, starting by the end of December during the winter observation periods 2015–2016 and 2016–2017. In the same paper, the snow depth reached its maximum of 26 cm on 30 January 2016 for the first winter and 99 cm on 7 April 2017 for the second winter. In this snow survey study, during the same two winters a snow depth of 27 cm was measured on 24 January 2016 and a snow depth of 86 cm on 27 March 2017. Furthermore, at an elevation of 4900 m a.s.l. in South Phulu, a snow depth of 79 cm was recorded with a terrestrial camera on 10 April 2017, using a snow stake. This also showed good agreement with the captured snow depth peaks, accumulation and ablation periods, with just minor variations. The winter of 2017–2018 showed the least snowpack accumulation and the SCA remained at a significant below average level up to the summer season in the Ladakh region. Similar reports by Dharpure et al. (2021) and Passang et al. (2022) revealed that 2017–2018 was the most snow-scarce winter over the whole study period from 2000 to 2019.

In a study by Tiwari et al. (2016), the snow water equivalent (SWE) over the period 1979–2007 was examined over the Western Himalayas, with special focus on the (Indian) Satluj basin area, where accumulated SWE values increased from an average of around 40 mm per month in December to around 90 mm in March, with extremes of more than 120 mm per month during some years. In general a large interannual variation in the SWE over the whole Western Himalayan region was observed.

During the short observation period (2015–2018) of the three snow plots in this research, SWE values approximating 40 mm showed only in January, and higher values were recorded in February (up to 98 mm) and March (a maximum of 128 mm).

4.6 Correlation of SCA and RSLE Against Meteorological Parameters

Pearson's r and Spearman's ρ correlation coefficient methods were implemented to find the relationship between the monthly mean SCA, Regional Snow Line Elevation (RSLE), Gyamtsa automatic weather station (AWS) air temperature, and Khakshal AWS air and soil temperatures, at significance levels of $p < 0.05$ and < 0.01.

The correlation analysis of the SCA and air and soil temperature in the periods from January 2016 to February 2017 (except for June and July 2016 for the Gyamtsa AWS, due to data gaps), and from August 2016 to September 2019 (except from February to April 2017 for the Khakshal AWS, due to data gaps) shows a significant negative correlation ($p < 0.05$ for Gyamtsa and $p < 0.01$ for Khakshal). These findings signify the influence of temperature on the SCA, where the SCA decreases with increasing temperature, and are in agreement with previous studies in the Western Himalayas and Karakoram (Viste and Sorteberg 2015; Kour et al. 2016; Dharpure et al. 2020, 2021; Thapa et al. 2020; Choudhury et al. 2021). For example, Dharpure et al. (2021), using MODIS snow cover products, MODIS Land Surface

Temperature products, and ERA5 climate data in the period 2000–2019, reported that the air and surface temperature show an inverse relation with the SCA over the Karakorum and Western Himalayan regions, with Pearson's coefficients of –0.2 to –0.8 for SCA against air temperature and –0.6 to –0.8 for SCA against surface temperature, at significance levels of $p < 0.05$ and < 0.001. Choudhury et al. (2021) investigated the SCA and its response to temperature over the north-west Himalaya during the same period, on the basis of MODIS Terra 8-day composite snow cover (MOD10A2) and daily land surface temperature data (MOD11A1). Spearman's correlation test showed statistically significant negative correlations of –0.81, –0.80, –0.45, –0.73, and –0.53 for, respectively, annual, pre-monsoon (Mar–May), monsoon (Jun–Sep), post-monsoon (Oct–Nov), and winter (Dec–Feb) seasons, with p-values between < 0.05 and 0.001.

The calculated correlation values of the SCA against temperature in this study fall in similar ranges, the air temperature varying between –0.69 (for Gyamtsa) and –0.79 (for Khakshal) with Pearson's test, and –0.72 (Gyamtsa) to –0.80 (Khakshal) with Spearman's test. Similarly, Pearson's and Spearman's methods show negative correlations of –0.75 and –0.78, respectively, for the soil temperature at Khakshal. It is important to note here that due to the shorter and incomplete observation period, it is challenging to compare the study findings with studies that analyse longer periods.

Further correlation analysis between the average RSLE and the same average temperatures during the same periods shows a significant positive correlation ($p < 0.01$) in the study area, which indicates the influence of temperature on the RSLE, with an upward shift of the RSLE with increasing temperature. This is also reflected in earlier studies (Tang et al. 2014, 2020; Viste and Sorteberg 2015; Kour et al. 2016). Kour et al. (2016) delineated from MODIS snow cover 8-day MOD09A1 data the SCA and RSLE (in the paper called LESCA: the lowest elevation from where snow cover area starts) for parts of the Chenab basin, Western Himalayas, in the observation period 2000–2012. The data were analysed against weather variables of meteorological stations, with the Batote station denoting a significant positive correlation between the RSLE and mean air temperatures, with for the months March and May Pearson's coefficients of 0.56 and 0.71, respectively, at a significance level of $p < 0.05$. In the month of December however, a significant negative correlation of –0.65 was observed at Reasi station. Similarly, Tang et al. (2014) used MOD10A1 of MODIS collection 5 and temperature data from 102 meteorological stations, to extract and evaluate snow line elevations on the Tibetan Plateau over the period from 2001 to 2013. In seven subregions significant Pearson's test correlations were observed from February to September, between 0.56 and 0.92 at $p < 0.05$ and < 0.01, of which May and June had the highest averages. Also, in a study about snow line altitudes at the end of the melting season, across the High Mountain Asia (HMA) region in the years 2001–2016, on the basis of MODIS MOD10A1 and ERA5 temperature data, Tang et al. (2020) reported for the summer season significant correlations with Pearson's r, with 0.64 ($p < 0.01$) in the Western Himalayas and 0.57 ($p < 0.05$) in the Karakoram.

In this research the RSLE against air temperature is only of a significant positive correlation, with values in the range from 0.84 (Gyamtsa) to 0.89 (Khakshal). Here too, due to the lack of sufficient data the results are not easily collated with other studies.

4.7 Validation and Accuracy Assessment

In the Himalayan context, many authors have used satellite, field observation, and meteorological data to assess the accuracy of the cloud removal approach for MODIS snow data (Jain et al. 2008; Dharpure et al. 2021; Muhammad and Thapa 2021). The threshold value of NDSI ≥ 0.40 applied at three level resolutions, from moderate MODIS to high Landsat 8 OLI, to very high resolution orthophotos snow data, agrees well to distinguish snow from no snow pixels in the study area. The same NDSI value was achieved in a study using IRS-AWiFS satellite data to monitor the seasonal snow cover of the Zangskar basin in the period 2004–2007 (Ajai et al. 2010), and in another study in the Satluj river basin (Jain et al. 2008), comparisons were made between IRS-WiFS, NOAA, and MODIS data in snow cover mapping under Himalayan conditions. An NDSI ≥ 0.40 threshold was applied for MODIS, and a good agreement with ground-based data was observed. Furthermore, Khali et al. (2022) in Zangskar Valley and numerous authors (Sood et al. 2020; Dharpure et al. 2021; Muhammad and Thapa 2021) used the global threshold of NDSI of 0.40 over the entire Himalayas.

The final output of MODIS cloud-free snow maps showed a good accuracy congruence level with high-resolution Landsat 8 OLI and very high-resolution time-lapse orthorectified photos at a small catchment scale in Leh Valley. Several studies in the past utilised time-lapse photographs in Europe (Parajka et al. 2012; Dong and Menzel 2017), which was also successfully applied in the South Phulu location to infer the settling characteristic of the snowpack and the spatio-temporal patterns of the snow cover over various landscape configurations. The achieved error results by the WSL Monoplotting Tool (MPT, developed at the Swiss Federal Institute for Forest, Snow, and Landscape Research) (Bozzini et al. 2011) showed for terrestrial camera station Gyamtsa mean 2D and 3D errors of, respectively, 6.2 m and 6.6 m, for Khakshal 2.9 m and 3.2 m, for Lungmar 8.7 m and 9.5 m, and for South Phulu 3.5 m and 3.8 m, which were slightly more than an example given by Steiner (2011) (mean 2D and 3D errors of 1.9 m and 2.2 m) and Stockdale (2017), who calculated average errors of 2.9 m for their best camera, 5 m for random-dispersed camera, and 9.1 m for worst camera. However, they had very high spatial resolution 1 m digital elevation models (DEMs) and up to 0.5 m orthophotos for their perusal. Errors in the MPT can be further mitigated combining time-lapse camera stations with drone photogrammetry and Light Detection and Ranging (LIDAR) solutions in future studies.

4.8 Some Reflections on the Socio-Hydrological Aspects of Snow Cover

Snow cover is integral to the global climate system, as it alters energy and moisture exchanges between the Earth's surface and the atmosphere and serves as a water reservoir in hydrological systems. Its high reflectivity, with an albedo of 0.8–0.9 for fresh snow, significantly influences the energy interaction between the surface and the atmosphere. Furthermore, snow cover is a crucial water source, providing water for agricultural irrigation and other uses (Thackeray et al. 2019).

The temporal occurrence of snow events in the Ladakh region has undergone a discernible alteration (Passang et al. 2022), which has a notable influence on the characteristics of the snowpack at the higher catchment areas. This in turn bears significant implications for the hydrological dynamics of the region, including the timing and magnitude of streamflow, as well as the overall availability of water for irrigation purposes.

Research highlights the socio-hydrological significance of snow cover in maintaining the ecological balance of the region and the need for sustainable water management practices to ensure the availability of water resources in Ladakh's cold-arid climate (Nüsser et al. 2012, 2019; Nüsser 2017). Sufficient melt water provision from glaciers or permanent snowfields in the higher catchment areas, or, if local topography allows direct extraction from the primary stream, is an essential requirement for local livelihoods and irrigated crop cultivation (Nüsser et al. 2012).

Land use in the region has long grappled with seasonal water scarcity, impacting both irrigation and domestic water supply, as documented by Dame and Mankelow (2010) and Nüsser and Baghel (2016). Addressing this challenge, artificial glaciers can be instrumental in diminishing the water shortage problems. They act as ice storage systems that thaw at the beginning of the farming season, thus filling the crucial void in water supply (Norphel and Tashi 2015; Nüsser and Baghel 2016; Nüsser et al. 2019).

4.9 Limitations

Snowboard: Due to budgetary reasons no automatic snow depth and SWE sensors were installed. Because of the short period of measurements (2015–2019), except for a basic descriptive, comparative, and graphical analysis, no statistical tests and analyses like regression and correlation could be meaningfully performed.

Snow plots and snow stake: As a consequence of constraints in budget, time, and manpower, the inaccessibility of the terrain during the winter period, and the exposure to weather elements, avalanches, and high altitude, only a limited snow plot research could be conducted on three fairly horizontal snow plots (at 4770, 4810, and 4860 m a.s.l.) in relatively close proximity of each other. More sloping terrain had to be avoided due to the increased avalanche danger and imminent physical harm as a result of the rough moraine terrain filled with boulders and debris. Moreover, during the measured period of 2015–2018, only up to 7 January 2017 could all plots be surveyed together, after which the two upper plots proved inaccessible due to the increased snow cover, avalanches, and unpredictable underlying terrain conditions. The snow stake at a higher elevation (4900 m a.s.l.) was not at all accessible between the onset of the snowfall and the end of spring of the measured winter periods during 2016–2018. Additionally, the second year the site turned out to be vandalised, with the snow stake uprooted and smashed. Since this already must have happened at the beginning of the second winter, no snow depth data were recorded. Therefore, since snow measurements could only be performed at four different elevations of the relief (three snow plot locations and the snow stake), also here except for a basic descriptive, comparative, and graphical analysis, any regression or correlation analysis would not have produced statistically meaningful results.

The Gyamtsa AWS was temporarily shut down from 5 p.m. on 17 December 2015 until 1 p.m. on 1 January 2017, to add in parallel an extra solar panel to the initial single solar panel set-up. This is in order to generate additional power and sustain the battery during periods of snowfall and extreme low temperatures. The battery box was now also wrapped in an insulating material and buried underground. Further improvements and testing of the operating system were also implemented. These modifications led to an unavailability of data during this period. From 2 June 2016 until 27 August 2016, sensor weather data were not logged due to a glitch in the software; however camera data were recorded. At 2 a.m. on 21 February 2017, absence of data occurred as a result of heavy snowfall, which had buried both the solar panels. An ensuing prolonged period of cloudy days resulted in a lack of power to sustain the battery reserve. Thereupon the station was removed from the site.

The Khakshal AWS showed unavailability of sensor weather data from 3:41 a.m. on 13 June 2016 until 3:11 p.m. on 10 July 2016, due to an irregular internet connection. It however kept recording camera data. There was again a period of absence of data from 2:23 a.m. on 2 September 2016 until 11:33 a.m. on 9 September 2016, due to several electrical power surges burning the electric wiring. From 11:11 p.m. on 8 February 2017 until 7:22 p.m. on 3 April 2017, data showed missing due to the optical fibre cable for the Internet on the Zoji La pass being swept away by an avalanche. As a result the Photon microcontroller (Particle 2016) was unable to transfer data. From 12:56 p.m. on 9 April 2017 until 11:56 p.m. on 31 May 2017, data were absent again due to an electrical power surge burning the fuse of the voltage inverter at the premises, shutting down the station. Replacing the Photon microcontroller with a self-soldered weather board, upgrading the voltage inverter, and improvements in the stability of the mains electricity power supply in Leh Valley led to a constant data supply from the station. Now, only occasional maintenance breaks cause missing data for very brief periods.

References

Ajai NS, Kulkarni AV, Sharma A, Bahuguna I, Rathore B, Singh S, Patel J (2010) Snow and glacier studies. Space Applications Centre, ISRO, Ahmedabad

Bhutiyani M (1999) Mass-balance studies on Siachen glacier in the Nubra valley, Karakoram Himalaya, India. J Glaciol 45(149):112–118. https://doi.org/10.3189/S0022143000003099

Bilal H, Chamhuri S, Mokhtar MB, Kanniah KD (2019) Recent snow cover variation in the Upper Indus Basin of Gilgit Baltistan, Hindukush Karakoram Himalaya. J Mt Sci 16(2):296–308. https://doi.org/10.1007/s11629-018-5201-3

Box GE, Jenkins GM, Reinsel GC (2008) Time series analysis: forecasting and control. Wiley, Hoboken. https://doi.org/10.1002/9781118619193

Bozzini C, Conedera M, Krebs P (2011) A new tool for obtaining cartographic georeferenced data from single oblique photos. In: Proceedings of the XXIIIrd international CIPA symposium, Prague. WSL Swiss Federal Research Institute, Bellinzona, vol 1216, p 6. https://www.wsl.ch/fileadmin/user_upload/WSL/Projekte/Evoluzione_paesaggio/Bozzini_CIPA2011_en.pdf

Choudhury A, Yadav AC, Bonafoni S (2021) A response of snow cover to the climate in the Northwest Himalaya (NWH) using satellite products. Remote Sens 13(4):655. https://doi.org/10.3390/rs13040655

Cohen J (1960) A coefficient of agreement for nominal scales. Educ Psychol Meas 20(1):37–46. https://doi.org/10.1177/001316446002000104

Dame J, Mankelow JS (2010) Stongde revisited: land-use change in central Zangskar. Erdkunde 64(4):355–370. https://doi.org/10.3112/erdkunde.2010.04.05

Dharpure JK, Patel A, Goswami A, Kulkarni AV, Snehmani (2020) Spatiotemporal snow cover characterization and its linkage with climate change over the Chenab river basin, western Himalayas. GIScience Remote Sens 57(7):882–906. https://doi.org/10.1080/15481603.2020.1821150

Dharpure JK, Goswami A, Patel A, Kulkarni AV, Snehmani (2021) Assessment of snow cover variability and its sensitivity to hydrometeorological factors in the Karakoram and Himalayan region. Hydrol Sci J 66(15):2198–2215. https://doi.org/10.1080/02626667.2021.1985125

Dong C, Menzel L (2016) Producing cloud-free MODIS snow cover products with conditional probability interpolation and meteorological data. Remote Sens Environ 186:439–451. https://doi.org/10.1016/j.rse.2016.09.019

Dong C, Menzel L (2017) Snow process monitoring in montane forests with time-lapse photography. Hydrol Process 31(16):2872–2886. https://doi.org/10.1002/hyp.11229

Fayaz M, Meraj G, Khader SA, Farooq M (2022) ARIMA and SPSS statistics based assessment of landslide occurrence in western Himalayas. Environ Challenges 9:100624. https://doi.org/10.1016/j.envc.2022.100624

Gafurov A, Bárdossy A (2009) Cloud removal methodology from MODIS snow cover product. Hydrol Earth Syst Sci 13(7):1361–1373. https://doi.org/10.5194/hess-13-1361-2009

Gao Y, Xie H, Yao T, Xue C (2010) Integrated assessment on multi-temporal and multi-sensor combinations for reducing cloud obscuration of MODIS snow cover products of the Pacific Northwest USA. Remote Sens Environ 114(8):1662–1675. https://doi.org/10.1016/j.rse.2010.02.017

Gascoin S, Hagolle O, Huc M, Jarlan L, Dejoux JF, Szczypta C, Marti R, Sánchez R (2015) A snow cover climatology for the Pyrenees from MODIS snow products. Hydrol Earth Syst Sci 19(5):2337–2351. https://doi.org/10.5194/hess-19-2337-2015

Gurung DR, Giriraj A, Aung KS, Shrestha BR, Kulkarni AV (2011) Snow-cover mapping and monitoring in the Hindu Kush-Himalayas. Technical report, International Centre for Integrated Mountain Development (ICIMOD), Kathmandu. https://doi.org/10.53055/ICIMOD.550

Gurung DR, Maharjan SB, Shrestha AB, Shrestha MS, Bajracharya SR, Murthy M (2017) Climate and topographic controls on snow cover dynamics in the Hindu Kush Himalaya. Int J Climatology 37(10):3873–3882. https://doi.org/10.1002/joc.4961

Hall DK, Riggs GA (2007) Accuracy assessment of the MODIS snow products. Hydrol Process 21(12):1534–1547. https://doi.org/10.1002/hyp.6715
Hall DK, Riggs GA, Salomonson VV (1995) Development of methods for mapping global snow cover using moderate resolution imaging spectroradiometer data. Remote Sens Environ 54(2):127–140. https://doi.org/10.1016/0034-4257(95)00137-P
Hall DK, Riggs GA, Salomonson VV, Barton J, Casey K, Chien J, DiGirolamo N, Klein A, Powell H, Tait A (2001) Algorithm theoretical basis document (ATBD) for the MODIS snow and sea ice-mapping algorithms. NASA/Goddard Space Flight Center 45:14–28. https://eospso.gsfc.nasa.gov/sites/default/files/atbd/atbd_mod10.pdf
Hall DK, Riggs GA, Salomonson VV, DiGirolamo NE, Bayr KJ (2002) MODIS snow-cover products. Remote Sens Environ 83(1–2):181–194. https://doi.org/10.1016/S0034-4257(02)00095-0
Hassan SA, Ansari MRK (2010) Nonlinear analysis of seasonality and stochasticity of the Indus River. Hydrol Sci J–Journal des Sciences Hydrologiques 55(2):250–265. https://doi.org/10.1080/02626660903546167
Hauksson BÞ (2017) Estimation of regional snow-line elevations in Norway. Testing a method for estimating RSLE and melt slope analysis in Norway. Master's thesis, University of Oslo, Oslo. http://urn.nb.no/URN:NBN:no-60729
Hyndman RJ, Athanasopoulos G (2021) Forecasting: principles and practice. OTexts, Melbourne. https://otexts.com/fpp3
Jain SK, Goswami A, Saraf AK (2008) Accuracy assessment of MODIS, NOAA and IRS data in snow cover mapping under Himalayan conditions. Int J Remote Sens 29(20):5863–5878. https://doi.org/10.1080/01431160801908129
Jain SK, Goswami A, Saraf AK (2009) Role of elevation and aspect in snow distribution in Western Himalaya. Water Res Manag 23(1):71–83. https://doi.org/10.1007/s11269-008-9265-5
Khali H, Pandey P, Ali SN, Champatiray PK (2022) MODIS observed spatiotemporal variation of snow cover in Zanskar Valley, North-Western Himalaya. Front Water 4:853146. https://doi.org/10.3389/frwa.2022.853146
Khan H, Hassan SA (2015) Stochastic river flow modelling and forecasting of upper Indus basin. J Basic Appl Sci 11:630–636. https://doi.org/10.6000/1927-5129.2015.11.84
Khan SI, Adams TE (2019) Indus River Basin: water security and sustainability. Elsevier, Amsterdam. https://doi.org/10.1016/C2016-0-04119-0
Klein AG, Barnett AC (2003) Validation of daily MODIS snow cover maps of the Upper Rio Grande River Basin for the 2000–2001 snow year. Remote Sens Environ 86(2):162–176. https://doi.org/10.1016/S0034-4257(03)00097-X
Kour R, Patel N, Krishna AP (2016) Assessment of temporal dynamics of snow cover and its validation with hydro-meteorological data in parts of Chenab Basin, western Himalayas. Sci China Earth Sci 59(5):1081–1094. https://doi.org/10.1007/s11430-015-5243-y
Krajčí P, Holko L, Perdigão RA, Parajka J (2014) Estimation of regional snowline elevation (RSLE) from MODIS images for seasonally snow covered mountain basins. J Hydrol 519:1769–1778. https://doi.org/10.1016/j.jhydrol.2014.08.064
Kumar M, Tiwari RK, Kumar K, Rautela KS (2022) Prediction of seasonal snow accumulation and depletion by SARIMA model using MODIS data. AQUA – Water Infrastructure, Ecosystems and Society, p 15. http://doi.org/10.21203/rs.3.rs-2024923/v1
Kumar M, Tiwari RK, Kumar K, Rautela KS (2023) Statistical evaluation of snow accumulation and depletion from remotely sensed MODIS snow time series data using the SARIMA model. J Water Supply Res Technol-Aqua 72(3):348–362. https://doi.org/10.2166/aqua.2023.231
Liang TG, Huang XD, Wu CX, Liu XY, Li WL, Guo ZG, Ren JZ (2008) An application of MODIS data to snow cover monitoring in a pastoral area: a case study in Northern Xinjiang, China. Remote Sens Environ 112(4):1514–1526. https://doi.org/10.1016/j.rse.2007.06.001
Muhammad S, Thapa A (2021) Daily Terra–Aqua MODIS cloud-free snow and Randolph Glacier Inventory 6.0 combined product (M* D10A1GL06) for high-mountain Asia between 2002 and 2019. Earth Syst Sci Data 13(2):767–776. https://doi.org/10.5194/essd-13-767-2021

Mukherjee A, Rakshit S, Nag A, Ray M, Kharbikar H, Shubha K, Sarkar S, Paul S, Roy S, Maity A et al (2016) Climate change risk perception, adaptation and mitigation strategy: an extension outlook in mountain Himalaya. In: Conservation agriculture: an approach to combat climate change in Indian Himalaya, chap 10. Springer, Singapore, pp 257–292. https://doi.org/10.1007/978-981-10-2558-7_10

Norphel C, Tashi P (2015) Snow water harvesting in the cold desert in Ladakh: an introduction to artificial glacier. In: Mountain hazards and disaster risk reduction. Springer, Tokyo, pp 199–210. https://doi.org/10.1007/978-4-431-55242-0_11

Nüsser M (2017) Socio-hydrology: a new perspective on mountain waterscapes at the Nexus of Natural and Social Processes. Mountain Res Dev 37(4):518–520. https://www.jstor.org/stable/90016621

Nüsser M, Baghel R (2016) Local knowledge and global concerns: artificial glaciers as a focus of environmental knowledge and development interventions. In: Meusburger P, Freytag T, Suarsana L (eds) Ethnic and cultural dimensions of knowledge. Knowledge and space, vol 8. Springer, Heidelberg, pp 191–209. https://doi.org/10.1007/978-3-319-21900-4_9

Nüsser M, Schmidt S, Dame J (2012) Irrigation and development in the Upper Indus Basin: characteristics and recent changes of a socio-hydrological system in Central Ladakh, India. Mountain Res Dev 32(1):51–61. https://doi.org/10.1659/MRD-JOURNAL-D-11-00091.1

Nüsser M, Dame J, Kraus B, Baghel R, Schmidt S (2019) Socio-hydrology of "artificial glaciers" in Ladakh, India: assessing adaptive strategies in a changing cryosphere. Reg Environ Change 19:1327–1337. https://doi.org/10.1007/s10113-018-1372-0

Parajka J, Blöschl G (2008) Spatio-temporal combination of MODIS images–potential for snow cover mapping. Water Res Res 44(3):13. https://doi.org/10.1029/2007WR006204

Parajka J, Pepe M, Rampini A, Rossi S, Blöschl G (2010) A regional snow-line method for estimating snow cover from MODIS during cloud cover. J Hydrol 381(3-4):203–212. https://doi.org/10.1016/j.jhydrol.2009.11.042

Parajka J, Holko L, Kostka Z, Blöschl G (2012) MODIS snow cover mapping accuracy in a small mountain catchment-comparison between open and forest sites. Hydrol Earth Syst Sci 16(7):2365. https://doi.org/10.5194/hess-16-2365-2012

Particle (2016) IoT platform. https://www.particle.io. Accessed 1 July 2016

Parvaze S, Khan JN, Kumar R, Allaie SP (2021) Temporal flood forecasting for trans-boundary Jhelum River of Greater Himalayas. Theor Appl Climatology 144:493–506. https://doi.org/10.1007/s00704-021-03562-8

Passang S, Schmidt S, Nüsser M (2022) Topographical impact on snow cover distribution in the Trans-Himalayan Region of Ladakh, India. Geosciences 12(8):311. https://doi.org/10.3390/geosciences12080311

Paudel KP, Andersen P (2011) Monitoring snow cover variability in an agropastoral area in the Trans Himalayan region of Nepal using MODIS data with improved cloud removal methodology. Remote Sens Environ 115(5):1234–1246. https://doi.org/10.1016/j.rse.2011.01.006

Rawat D, Mishra P, Ray S, Warnakulasooriya HHF, Sati SP, Mishra G, Alkattan H, Abotaleb M (2022) Modeling of rainfall time series using NAR and ARIMA model over western Himalaya, India. Arab J Geosci 15(23):1696. https://doi.org/10.1007/s12517-022-10994-7

Salomonson VV, Appel I (2004) Estimating fractional snow cover from MODIS using the normalized difference snow index. Remote Sensing of Environment 89(3):351–360. https://doi.org/10.1016/j.rse.2003.10.016

Saydi M, Ding Jl (2020) Impacts of topographic factors on regional snow cover characteristics. Water Sci Eng 13(3):171–180. https://doi.org/10.1016/j.wse.2020.09.002

Schmidt S, Nüsser M (2012) Changes of high altitude glaciers from 1969 to 2010 in the Trans-Himalayan Kang Yatze Massif, Ladakh, northwest India. Arct Antarct Alp Res 44(1):107–121. https://doi.org/10.1657/1938-4246-44.1.107

Singh D, Juyal V, Sharma V (2016) Consistent seasonal snow cover depth and duration variability over the Western Himalayas (WH). J Earth Syst Sci 125(7):1451–1461. https://doi.org/10.1007/s12040-016-0737-3

Sood V, Singh S, Taloor AK, Prashar S, Kaur R (2020) Monitoring and mapping of snow cover variability using topographically derived NDSI model over north Indian Himalayas during the period 2008–2019. Appl Comput Geosci 8:100040. https://doi.org/10.1016/j.acags.2020.100040

Steiner L (2011) Reconstruction of glacier states from geo-referenced, historical postcards. Master's thesis, Institut für Kartografie und Geoinformation, Eidgenössische Technische Hochschule Zürich, Zürich. https://www.geometh-data.ethz.ch/downloads/Steiner_Masterarbeit.pdf

Stockdale CA (2017) A century of landscape change in the southern Rocky Mountains and foothills of Alberta: using historical photography to quantify ecological change. PhD thesis, University of Alberta, Department of Renewable Resources, Edmonton. https://doi.org/10.7939/R3W37M744

Tahir AA, Chevallier P, Arnaud Y, Ashraf M, Bhatti MT (2015) Snow cover trend and hydrological characteristics of the Astore River basin (Western Himalayas) and its comparison to the Hunza basin (Karakoram region). Sci Total Environ 505:748–761. https://doi.org/10.1016/j.scitotenv.2014.10.065

Tang Z, Wang J, Li H, Liang J, Li C, Wang X (2014) Extraction and assessment of snowline altitude over the Tibetan plateau using MODIS fractional snow cover data (2001 to 2013). J Appl Remote Sens 8(1):084689–084689. https://doi.org/10.1117/1.JRS.8.084689

Tang Z, Wang X, Deng G, Wang X, Jiang Z, Sang G (2020) Spatiotemporal variation of snowline altitude at the end of melting season across High Mountain Asia, using MODIS snow cover product. Adv Space Res 66(11):2629–2645. https://doi.org/10.1016/j.asr.2020.09.035

Thackeray CW, Derksen C, Fletcher CG, Hall A (2019) Snow and climate: feedbacks, drivers, and indices of change. Curr Clim Change Rep 5:322–333. https://doi.org/10.1007/s40641-019-00143-w

Thapa S, Li B, Fu D, Shi X, Tang B, Qi H, Wang K (2020) Trend analysis of climatic variables and their relation to snow cover and water availability in the Central Himalayas: a case study of Langtang Basin, Nepal. Theor Appl Climatology 140(3):891–903. https://doi.org/10.1007/s00704-020-03096-5

Thenkabail PS (2015) Remote sensing of water resources, disasters, and urban studies. CRC Press, Boca Raton. https://doi.org/10.1201/b19321

Tiwari S, Kar SC, Bhatla R (2016) Interannual variability of snow water equivalent (SWE) over Western Himalayas. Pure Appl Geophys 173:1317–1335. https://doi.org/10.1007/s00024-015-1163-1

Tran H, Nguyen P, Ombadi M, Hsu Kl, Sorooshian S, Qing X (2019) A cloud-free MODIS snow cover dataset for the contiguous United States from 2000 to 2017. Sci Data 6:180300. https://doi.org/10.1038/sdata.2018.300

Viste E, Sorteberg A (2015) Snowfall in the Himalayas: an uncertain future from a little-known past. The Cryosphere 9(3):1147–1167. https://doi.org/10.5194/tc-9-1147-2015

Wang J (2017) Satellite observation on snow. In: Training workshop on water resource management and sustainable development strategies in arid regions of Central/Western Asia under climate change 2017, Northwest Institute of Eco-Environment and Resources, Chinese Academy of Sciences, Lanzhou

Wani JM, Thayyen RJ, Ojha CSP, Gruber S (2021) The surface energy balance in a cold and arid permafrost environment, Ladakh, Himalayas, India. The Cryosphere 15(5):2273–2293. https://doi.org/10.5194/tc-15-2273-2021

Xu L, Niu R, Penga L (2008) Snow cover area changing analysis over Tibetan Plateau with MODIS and ASTER data. In: Proceedings of ISPRS congress, Beijing. International Society for Photogrammetry and Remote Sensing, Hannover. https://www.isprs.org/proceedings/XXXVII/congress/7_pdf/10_ThS-18/15.pdf

Zhang Y, Yan S, Lu Y (2010) Snow cover monitoring using MODIS data in Liaoning Province, Northeastern China. Remote Sens 2(3):777–793. https://doi.org/10.3390/rs2030777

Zhang Y, Cao T, Kan X, Wang J, Tian W (2017a) Spatial and temporal variation analysis of snow cover using MODIS over Qinghai-Tibetan Plateau during 2003–2014. J Indian Soc Remote Sens 45(5):887–897. https://doi.org/10.1007/s12524-016-0617-y

Zhang Y, Kan X, Ren W, Cao T, Tian W, Wang J (2017b) Snow cover monitoring in Qinghai-Tibetan Plateau based on Chinese Fengyun-3/VIRR data. J Indian Soc Remote Sens 45(2):271–283. https://doi.org/10.1007/s12524-015-0527-4

Zhang H, Zhang F, Zhang G, Che T, Yan W, Ye M, Ma N (2019) Ground-based evaluation of MODIS snow cover product V6 across China: implications for the selection of NDSI threshold. Sci Total Environ 651:2712–2726. https://doi.org/10.1016/j.scitotenv.2018.10.128

Chapter 5
Synthesis and Insights Charting the Future of Cryospheric Research in Ladakh

Abstract This chapter is split into two parts: conclusions and outlook. The conclusions synthesise the results, exploring the seasonal snow cover in Ladakh's Trans-Himalayan region, assessing its distribution and dynamics through advanced methodologies. The calibrated Normalised Difference Snow Index (NDSI) and cloud reduction algorithms for MODIS imagery provide a refined understanding of snow cover area (SCA) and snow cover days (SCD). Field surveys and automatic weather station (AWS) data offer insights into the topographical and climatic influences on snow patterns. Despite slight, non-significant trends in SCA and SCD, the research highlights the importance of snow cover for water resource management and climate change adaptation. Integrating terrestrial and satellite imagery, the study enhances snow mapping accuracy and underscores the need for localised, detailed investigations. The outlook propounds the future research direction, including deploying high-resolution data and time-lapse cameras for nuanced, subregional analyses and maintaining Sentinel-3 and MODIS data continuity to monitor environmental changes. These efforts are vital for sustainable development and informed decision-making in high-altitude regions facing climatic challenges.

Keywords Snow · MODIS · NDSI · Outlook · Automatic weather station · Statistical analysis · Ladakh · India

5.1 Conclusions

Snow cover is a crucial component of high-altitude ecosystems and hydrological systems, playing a pivotal role in regulating climate, water resources, and ecological processes. The comprehensive analysis presented in this book delves into various facets of snow cover dynamics in the Trans-Himalayan region of Ladakh, offering valuable insights into the complexities of snow distribution, seasonal variations, topographical impacts, and the potential influence of climate change. The scientific investigations span multiple methodologies, including the calibration of the Normalised Difference Snow Index (NDSI), cloud reduction algorithms for Moderate Resolution Imaging Spectroradiometer (MODIS) satellite imagery, analyses of

S. Passang, *Snow Cover Distribution and Dynamics*, Advances in Asian Human-Environmental Research, https://doi.org/10.1007/978-3-031-57692-8_5

snow cover area (SCA) and snow cover days (SCD), influence of the relief on snow cover distribution, and the development of an Autoregressive Integrated Moving Average (ARIMA) model for snowfall forecasting. Additionally, this study explores field surveys and meteorological data collected from automatic weather stations (AWSs) and presents a comparative analysis of snow lines from north-facing and south-facing aspects delineated from terrestrial orthorectified photos.

The mountainous region of Ladakh contain vital natural resources of cryospheric fresh water in the form of snow, glaciers, and permafrost and act as a water reservoir for the many rivers originating in it. The snow cover distribution in the region plays especially an important and unique role in the sustenance of key components of the cryosphere, which ensure availability of water in the streams during the melt periods, for irrigating agricultural fields and downstream settlements. Thus, it offers baseline information and insights into how the water resources should be managed in the long term in the high mountain regions.

This study illustrates the potential of integrating terrestrial imagery with satellite imagery, using the monophotogrammetry principle, which allows to more accurately achieve the multi-resolution observation assessment of snow cover characteristics at various scales and time. The research emphasises the significance of NDSI threshold calibration in remote sensing applications for accurate snow cover mapping, particularly in regions characterised by diverse terrain and varying climatic conditions like Ladakh. Through rigorous analysis and comparison of MODIS and Landsat snow cover data, an optimal NDSI threshold value of 0.40, which also represents the global NDSI threshold, with an overall accuracy of 87.20%, was identified for Ladakh. This threshold exhibits spatial and seasonal variability, with winter and spring consistently exceeding 0.40, while summer and autumn exhibit lower values. The findings provide critical insights into Ladakh's specific NDSI requirements, enhancing the precision of snow cover assessments in the region.

Furthermore, the results fit well when comparisons are made amongst coarse MODIS, high-resolution Landsat 8 Operational Land Imager (LC8 OLI), and terrestrial oblique photos of a very high resolution.

The proposed algorithms for the reduction of the cloud obscuration of MODIS snow cover maps, encompassing Terra-Aqua combination, temporal combination, and spatio-temporal combination methods, prove effective in reducing cloud cover. The resulting cloud-free snow cover product opens new avenues for hydrological modelling, agriculture, and environmental studies, emphasising the importance of reliable and accurate satellite imagery in these domains.

The analysis of SCA and SCD in the study area reveals distinct seasonal patterns and subtle trends in snow cover dynamics. Although annual trends are generally non-significant, subregional variations highlight specific areas experiencing changes in snow cover extent and duration. The trend results demonstrate for the entire study area a slight non-significant increase in trend of the SCA in summer and annual, at rates of 0.08 and 0.2% year^{-1}, respectively, and a slight non-significant decrease in trend in winter and spring. Annually only 6% of all SCD pixels show a significant positive trend, and 2% of all pixels show a significant negative trend. A significant positive trend indicates a downward shift of the Regional Snow Line Elevation (RSLE) in summer (–3.53 m year^{-1}).

The seasonal snow cover across seven cold-arid subregions of Ladakh over the study period of almost two decades shows a high intra-annual and inter-annual variability. The results of the trend analysis show that significant increasing pixel trends with p-value < 0.1 are located in Suru and Zangskar, and significant decreasing trends in winter are situated mainly in the eastern part of Ladakh, in the Changthang area. These findings are instrumental in monitoring and managing water resources in regions heavily reliant on snowmelt for various purposes.

A comprehensive assessment of the topographical influence on snow cover distribution underscores the strong correlation between elevation, slope, and aspect with snow cover patterns. Higher elevations consistently maintain snow cover, while slopes and aspects influence snow distribution. These findings have significant implications for understanding snow cover dynamics, especially in high-altitude regions where topography plays a crucial role in shaping snow accumulation and persistence.

The Seasonal Autoregressive Integrated Moving Average (SARIMA) model in this research shows a promising potential to represent future snowfall in the region of Ladakh and test values of a forecast of 4 years, tested against the actual MODIS snow data demonstrated a good agreement, with an obtained coefficient of determination (R^2) value of 0.78.

Electronic custom-built outdoor AWSs, using different multi-configurable environmental sensors, were field-tested at the Gyamtsa and Khakshal locations. The Gyamtsa air temperature and Khakshal air and soil temperature variables proved useful in a correlation analysis with the MODIS monthly mean SCA and RSLE. Tests were carried out at a significance level of p-value < 0.05 for the Gyamtsa and p-value < 0.01 for the Khakshal AWSs. The observed correlations between temperature, SCA, and RSLE highlight the potential consequences of climate change in high-altitude regions.

The comparative analysis of snow lines from south-facing (Gyamtsa) and north-facing (Lungmar) aspects underscores the significant differences in snow cover persistence and elevation. North-facing slopes maintain snow cover for more extended periods at lower elevations, while south-facing slopes experience quicker snowmelt and snow lines are at higher elevations than north-facing slopes. This comparative analysis emphasises the importance of slope orientation in influencing local snow cover dynamics and its broader implications for hydrology, ecosystems, and climate studies.

Overall, this study underlines the relevance of regional and subregional analyses to capture the spatio-temporal dynamics of seasonal snow cover, and the results highlight the potential of detailed investigations of the snow cover combining terrestrial imagery with satellite imagery at local and subregional scales, in order to obtain more robust information. The findings contribute to our understanding of the intricate interplay between climatic factors, local topography, and snowpack dynamics in high-altitude regions. These insights are essential for monitoring and managing water resources, assessing climate change impacts, and making informed decisions regarding resource management and adaptation strategies in these challenging environments. The methodologies presented, including NDSI threshold calibration, cloud reduction algorithms, ARIMA modelling, field surveys, AWS data analysis, and snow line comparisons, collectively advance our knowledge of snow cover dynamics and their implications for a wide range of scientific disciplines and practical applications.

5.2 Outlook

This scientific investigation establishes a robust groundwork for future research endeavours in the region, promoting ongoing surveillance and data gathering to attain deeper insights into the evolving dynamics of snow cover in response to shifting climatic conditions. The comprehension of snow coverage in high-altitude locales such as Ladakh remains essential in addressing contemporary challenges related to the management of water resources, adaptation to climate change, preservation of ecosystems, and sustainable development.

To gain a more comprehensive understanding of snow cover dynamics, conducting a multi-subregional study at a highly localised scale becomes imperative. This entails the deployment of multiple terrestrial time-lapse camera stations within each subregion, complemented by the acquisition of high-resolution remote sensing and drone data. These measures will be instrumental in future snow research efforts.

To bolster the monitoring of snow cover, it is crucial to maintain the continuity of, e.g., Sentinel-3 data in conjunction with the existing MODIS snow data. Additionally, the incorporation of MODIS Land Surface Temperature data to monitor surface temperature fluctuations, Noah-MP Land Surface Model data to uphold a consistent overview of surface energy and water balance, and ERA5 data for quantifying long-term monthly total precipitation and air temperature trends will assume a pivotal role in assessing the impact of changing climatic conditions on the environment and ecosystem of the study area.

Appendix A
NDSI Threshold Values

Table A.1 NDSI threshold values. The highest kappa coefficient and corresponding NDSI values, accuracy, and percentage of snow are highlighted in red

NDSI	Accuracy %	Kappa coefficient	Snow %
0.40	87.45	0.73807	39.96
0.39	87.43	0.73812	40.49
0.38	87.39	0.73802	41.05
0.37	87.36	0.73808	41.65
0.36	87.30	0.73749	42.18
0.35	87.20	0.73609	42.73
0.34	87.09	0.73426	43.17
0.33	86.99	0.73272	43.63
0.32	86.89	0.73132	44.19
0.31	86.73	0.72867	44.70
0.30	86.58	0.72620	45.20
(Values of Landsat 8 northern tile vs MODIS Terra (Spring, 31 March 2018))			
0.47	85.21	0.70430	51.06
0.46	85.22	0.70454	51.67
0.45	85.21	0.70452	52.26
0.44	85.15	0.70339	52.84
0.43	85.06	0.70169	53.40
0.42	84.99	0.70038	53.95
0.41	84.90	0.69856	54.45
0.40	84.78	0.69614	55.03
0.39	84.65	0.69371	55.61
0.38	84.56	0.69201	56.08
0.37	84.45	0.68974	56.59

(continued)

S. Passang, *Snow Cover Distribution and Dynamics*, Advances in Asian Human-Environmental Research, https://doi.org/10.1007/978-3-031-57692-8

Table A.1 (continued)

NDSI	Accuracy %	Kappa coefficient	Snow %
0.36	84.28	0.68652	57.06
0.35	84.11	0.68321	57.58
0.34	83.95	0.68007	58.14
0.33	83.79	0.67694	58.64
0.32	83.61	0.67341	59.20
0.31	83.39	0.66912	59.73
0.30	83.17	0.66476	60.26
(Values of Landsat 8 southern tile vs MODIS Terra (Spring, 31 March 2018))			
0.40	86.62	0.71502	35.42
0.39	86.65	0.71611	35.74
0.38	86.68	0.71720	36.09
0.37	86.70	0.71792	36.43
0.36	86.69	0.71827	36.79
0.35	86.70	0.71879	37.12
0.34	86.68	0.71889	37.47
0.33	86.64	0.71848	37.85
0.32	86.61	0.71824	38.20
0.31	86.55	0.71755	38.55
0.30	86.51	0.71714	38.90
(Values of Landsat 8 northern tile vs MODIS Aqua (Spring, 31 March 2018))			
0.43	85.57	0.71120	47.82
0.42	85.59	0.71157	48.23
0.41	85.56	0.71102	48.61
0.40	85.55	0.71084	49.03
0.39	85.53	0.71046	49.42
0.38	85.50	0.70994	49.78
0.37	85.49	0.70990	50.18
0.36	85.46	0.70918	50.58
0.35	85.39	0.70801	50.96
0.34	85.35	0.70710	51.31
0.33	85.25	0.70520	51.66
0.32	86.17	0.70361	52.04
0.31	85.11	0.70248	52.40
0.30	85.04	0.70118	52.72
(Values of Landsat 8 southern tile vs MODIS Aqua (Spring, 31 March 2018))			
0.40	91.29	0.69688	18.16
0.39	91.26	0.69745	18.39
0.38	91.16	0.69592	18.64
0.37	91.06	0.69459	18.94
0.36	90.97	0.69370	19.24
0.35	90.87	0.69255	19.60
0.34	90.77	0.69118	19.86

(continued)

Table A.1 (continued)

NDSI	Accuracy %	Kappa coefficient	Snow %
0.33	90.62	0.68831	20.17
0.32	90.50	0.68647	20.50
0.31	90.39	0.68508	20.83
0.30	90.25	0.68288	21.15
(Values of Landsat 8 northern tile vs MODIS Terra (Summer, 20 September 2017))			
0.40	92.89	0.54305	8.85
0.39	92.81	0.54332	9.06
0.38	92.73	0.54496	9.34
0.37	92.63	0.54474	9.60
0.36	92.54	0.54544	9.86
0.35	92.43	0.54506	10.13
0.34	92.36	0.54596	10.35
0.33	92.24	0.54479	10.63
0.32	92.14	0.54509	10.92
0.31	91.99	0.54367	11.24
0.30	91.86	0.54279	11.54
(Values of Landsat 8 southern tile vs MODIS Terra (Summer, 20 September 2017))			
0.40	92.19	0.72214	17.20
0.39	92.15	0.72196	17.41
0.38	92.11	0.72205	17.61
0.37	92.08	0.72195	17.79
0.36	92.04	0.72210	18.00
0.35	92.02	0.72242	18.18
0.34	91.98	0.72244	18.38
0.33	91.96	0.72301	18.57
0.32	91.94	0.72357	18.77
0.31	91.90	0.72314	18.94
0.30	91.83	0.72220	19.15
(Values of Landsat 8 northern tile vs MODIS Aqua (Summer, 20 September 2017))			
0.40	93.87	0.60339	8.72
0.39	93.82	0.60423	8.91
0.38	93.78	0.60566	9.12
0.37	93.73	0.60621	9.29
0.36	93.66	0.60633	9.50
0.35	93.60	0.60629	9.69
0.34	93.54	0.60668	9.89
0.33	93.49	0.60787	10.09
0.32	93.42	0.60738	10.30
0.31	93.34	0.60701	10.51
0.30	93.24	0.60535	10.72

(continued)

Table A.1 (continued)

NDSI	Accuracy %	Kappa coefficient	Snow %
(Values of Landsat 8 southern tile vs MODIS Aqua (Summer, 20 September 2017))			
0.40	83.54	0.67081	52.29
0.39	83.54	0.67086	53.25
0.38	83.54	0.67094	54.21
0.37	83.55	0.67105	55.11
0.36	83.44	0.66898	55.97
0.35	83.34	0.66692	56.83
0.34	83.16	0.66348	57.62
0.33	83.95	0.65915	58.59
0.32	82.75	0.65533	59.32
0.31	82.49	0.65003	60.21
0.30	82.22	0.64473	60.99
(Values of Landsat 8 northern tile vs MODIS Terra (Autumn, 26 November 2018))			
0.46	81.05	0.62008	55.55
0.45	81.11	0.62117	56.39
0.44	81.05	0.61990	57.30
0.43	81.05	0.61959	58.17
0.42	80.97	0.61794	58.94
0.41	80.90	0.61642	59.69
0.40	80.81	0.61450	60.39
0.39	80.74	0.61290	61.20
0.38	80.61	0.61012	61.90
0.37	80.44	0.60658	62.60
0.36	80.24	0.60249	63.33
0.35	80.03	0.59826	63.98
0.34	79.80	0.59333	64.61
0.33	79.50	0.58726	65.32
0.32	79.29	0.58288	65.91
0.31	78.97	0.57641	66.62
0.30	78.70	0.57079	67.23
(Values of Landsat 8 southern tile vs MODIS Terra (Autumn, 26 November 2018))			
0.40	83.02	0.66042	51.28
0.39	83.08	0.66161	51.91
0.38	83.10	0.66216	52.57
0.37	83.14	0.66289	53.22
0.36	83.15	0.66304	53.83
0.35	83.09	0.66199	54.43
0.34	83.08	0.66180	54.99
0.33	83.04	0.66102	55.64
0.32	83.01	0.66038	56.16
0.31	82.95	0.65921	56.72
0.30	82.88	0.65771	57.31

(continued)

Table A.1 (continued)

NDSI	Accuracy %	Kappa coefficient	Snow %
(Values of Landsat 8 northern tile vs MODIS Aqua (Autumn, 26 November 2018))			
0.46	82.56	0.65049	55.08
0.45	82.57	0.65060	55.61
0.44	82.54	0.64981	56.08
0.43	82.51	0.65424	56.57
0.42	82.48	0.64861	57.09
0.41	82.44	0.64761	57.58
0.40	82.40	0.64684	58.06
0.39	82.37	0.64612	58.52
0.38	82.32	0.64503	58.97
0.37	82.26	0.64371	59.40
0.36	82.16	0.64177	59.88
0.35	82.07	0.63982	60.32
0.34	81.97	0.63779	60.76
0.33	81.87	0.63562	61.22
0.32	81.76	0.63341	61.64
0.31	81.63	0.63073	62.06
0.30	81.50	0.62799	62.50
(Values of Landsat 8 southern tile vs MODIS Aqua (Autumn, 26 November 2018))			
0.45	87.24	0.59841	82.20
0.44	87.42	0.59906	82.88
0.43	87.52	0.59696	83.60
0.42	87.65	0.59721	84.14
0.41	87.67	0.59420	84.63
0.40	87.71	0.59211	85.06
0.39	87.65	0.58673	85.51
0.38	87.61	0.58234	85.91
0.37	87.59	0.57798	86.35
0.36	87.52	0.57220	86.76
0.35	87.47	0.56772	87.08
0.34	87.40	0.56252	87.42
0.33	87.35	0.55777	87.77
0.32	87.28	0.55262	88.09
0.31	87.20	0.54676	88.40
0.30	87.08	0.53986	88.71
(Values of Landsat 8 northern tile vs MODIS Terra (Winter, 24 February 2017))			
0.50	90.72	0.60144	87.33
0.49	90.78	0.60404	87.75
0.48	90.85	0.60202	88.15
0.47	90.91	0.60070	88.48
0.46	90.95	0.59820	88.83

(continued)

Table A.1 (continued)

NDSI	Accuracy %	Kappa coefficient	Snow %
0.45	90.99	0.59591	89.13
0.44	91.00	0.59221	89.44
0.43	91.00	0.58909	89.71
0.42	90.99	0.58475	89.98
0.41	90.97	0.58030	90.23
0.40	90.97	0.57795	90.41
0.39	90.93	0.57286	90.64
0.38	90.89	0.56761	90.86
0.37	90.83	0.56134	91.11
0.36	90.79	0.55650	91.30
0.35	90.73	0.55014	91.53
0.34	90.66	0.54292	91.76
0.33	90.60	0.53704	91.94
0.32	90.53	0.53070	92.11
0.31	90.46	0.52494	92.28
0.30	90.38	0.51700	92.48
(Values of Landsat 8 southern tile vs MODIS Terra (Winter, 24 February 2017))			
0.44	85.52	0.57594	78.00
0.43	85.66	0.57672	78.47
0.42	85.72	0.57555	78.88
0.41	85.81	0.57509	79.31
0.40	85.88	0.57435	79.69
0.39	85.92	0.57308	80.05
0.38	85.95	0.57137	80.41
0.37	85.96	0.56923	80.75
0.36	86.01	0.56780	81.10
0.35	86.02	0.56569	81.42
0.34	85.99	0.56229	81.74
0.33	85.98	0.55966	82.04
0.32	85.95	0.55655	82.34
0.31	85.93	0.55380	82.58
0.30	85.91	0.55138	82.84
(Values of Landsat 8 northern tile vs MODIS Aqua (Winter, 24 February 2017))			
0.50	86.82	0.52674	81.28
0.49	86.93	0.52678	81.61
0.48	87.03	0.52642	81.94
0.47	87.09	0.52499	82.24
0.46	87.14	0.52321	82.53
0.45	87.19	0.52198	82.78
0.44	87.24	0.52038	83.04
0.43	87.26	0.51775	83.31

(continued)

Table A.1 (continued)

NDSI	Accuracy %	Kappa coefficient	Snow %
0.42	87.29	0.51569	83.55
0.41	87.31	0.51336	83.80
0.40	87.32	0.51080	84.03
0.39	87.33	0.50838	84.23
0.38	87.31	0.50464	84.46
0.37	87.31	0.50175	84.66
0.36	87.32	0.49902	84.87
0.35	87.29	0.49512	85.07
0.34	87.27	0.49150	85.25
0.33	87.25	0.48794	85.44
0.32	87.20	0.48343	85.63
0.31	87.17	0.47988	85.78
0.30	87.14	0.47574	85.98

(Values of Landsat 8 southern tile vs MODIS Aqua (Winter, 24 February 2017))

Appendix B
Snowboard Data

Table B.1 Snow depth, SWE, and snow density of the Khakshal snowboard in the period of 2015–2019

Date	HS_{RPi} cm	HS_{Tube} cm	SWE mm	ρ_S g cm^{-3}
11/11/2015		1.0		
27/11/2015		0.4		
11/12/2015		0.6		
25/12/2015		2.2	1	0.05
27/12/2015		3.0	2	0.07
04/01/2016		1.0		
05/01/2016		4.6	3	0.07
13/01/2016		1.4	1	0.07
26/01/2016		3.8	3	0.08
27/01/2016		3.8	3	0.08
29/01/2016		1.0		
12/02/2016		1.8		
02/01/2017	0.2	0.2		
08/01/2017	0.2	0.2		
16/01/2017	2.0	2.0	2	0.10
17/01/2017	0.9			
18/01/2017	0.9			
25/01/2017	0.3			
26/01/2017	0.5			
29/01/2017	1.2	1.2	2	0.17
30/01/2017	3.8			
31/01/2017	5.4	4.0	7	0.18
05/02/2017	5.9	4.6	5	0.11

(continued)

S. Passang, *Snow Cover Distribution and Dynamics*, Advances in Asian Human-Environmental Research, https://doi.org/10.1007/978-3-031-57692-8

Table B.1 (continued)

Date	HS_{RPi} cm	HS_{Tube} cm	SWE mm	ρ_S g cm^{-3}
06/02/2017	2.2	2.0	2	0.10
08/02/2017	3.2			
09/02/2017	4.5	2.2	3	0.14
13/02/2017		1.0		
19/02/2017	0.3			
20/02/2017	0.5			
21/02/2017	2.7			
01/03/2017	2.5			
08/03/2017	0.8			
09/03/2017	1.0			
05/04/2017	6.0			
06/04/2017	10.2	10.0	13	0.13
07/04/2017	7.0			
09/04/2017	1.6			
10/04/2017	2.5			
29/11/2017	0.6	0.6		
13/12/2017	0.9			
14/12/2017	0.8	0.8		
22/12/2017	0.3			
25/12/2017	0.7			
28/12/2017	0.2			
05/01/2018	0.1			
12/02/2018	0.7			
13/02/2018	1.1	1.2	2	0.17
21/04/2018	0.5	0.4		
08/10/2018	3.5			
04/11/2018	2.0			
14/11/2018	1.0			
19/11/2018	0.1			
14/12/2018	1.0			
05/01/2019	0.1			
07/01/2019	0.4			
08/01/2019	1.2			
09/01/2019	3.5			
13/01/2019	1.0			
16/01/2019	1.3			
17/01/2019	1.8			
19/01/2019	2.8			

(continued)

Table B.1 (continued)

Date	HS_{RPi} cm	HS_{Tube} cm	SWE mm	ρ_S g cm^{-3}
30/01/2019	0.2			
04/03/2019	12.0			
15/03/2019	5.2			
26/04/2019	0.5			

Khakshal snowboard: Longitude 77°58975'E, latitude 34°18346'N, and elevation 3660 m a.s.l.
HS_{RPi}: Snow depth measured with the RPi camera
HS_{Tube}: Snow depth measured with the snow tube
SWE: Snow water equivalent measured with the snow spring balance
ρ_S: Snow density

Appendix C
Snow Plot Data

Table C.1 Snow depth, SWE, and snow density of the three snow plots (2015–2018)

Date	Snow Plot	n	$\overline{HS}$ cm	SD cm	HS cm	SWE mm	$\overline{SWE}$ mm	ρ_S g cm^{-3}
10/12/2015	3	79	10	3.33	10	13	12	0.12
23/12/2015	1	30	11	2.75	9	14	17	0.15
	2	30	12	3.16	9	14	18	0.15
	3	26	15	3.85	10	15	22	0.14
12/01/2016	1	30	19	5.36	20	29	27	0.15
	2	32	18	2.66	16	26	28	0.16
	3	30	26	4.70	24	36	39	0.15
24/01/2016	1	35	23	6.08	27	45	39	0.17
	2	29	20	3.38	19	29	31	0.15
02/02/2016	1	34	42	4.84	48	67	59	0.14
	2	31	39	4.65	39	55	55	0.14
	3	31	48	5.43	54	86	76	0.16
25/02/2016	1	35	45	8.59	44	76	78	0.17
	2	34	40	5.01	36	55	61	0.15
	3	29	50	6.71	62	101	81	0.16
07/01/2017	1	35	10	1.74	7	12	18	0.17
	2	32	10	1.30	8	11	14	0.14
	3	34	11	1.82	11	19	19	0.17
01/02/2017	1	35	52	6.59	56	77	71	0.14
28/02/2017	1	35	61	6.78	68	110	98	0.16
27/03/2017	1	35	79	3.82	86	139	128	0.16
28/12/2017	1	35	15	1.52	12	18	22	0.15

(continued)

S. Passang, *Snow Cover Distribution and Dynamics*, Advances in Asian Human-Environmental Research, https://doi.org/10.1007/978-3-031-57692-8

Table C.1 (continued)

Date	Snow Plot	n	$\overline{HS}$ cm	SD cm	HS cm	SWE mm	$\overline{SWE}$ mm	ρ_S g cm^{-3}
08/01/2018	1	35	16	2.52	14	22	24	0.16
27/01/2018	1	35	12	4.02	14	22	19	0.16
02/02/2018	1	35	9	3.61	9	14	13	0.14
07/03/2018	1	35	17	7.05	9	15	30	0.17

Snow Plot 1: Longitude 77°61956'E, latitude 34°25937'N, and elevation 4770 m a.s.l.
Snow Plot 2: Longitude 77°61596'E, latitude 34°26080'N, and elevation 4810 m a.s.l.
Snow Plot 3: Longitude 77°61227'E, latitude 34°26409'N, and elevation 4860 m a.s.l.
n: Total snow depth measurements
$\overline{HS}$: Average snow depth
SD: Standard deviation of average snow depth
HS: Snow depth of snow sample
SWE: SWE of snow sample
$\overline{SWE}$: Average SWE
ρ_S: Snow density

Appendix D
Selected Trends Data

S. Passang, *Snow Cover Distribution and Dynamics*, Advances in Asian Human-Environmental Research, https://doi.org/10.1007/978-3-031-57692-8

Table D.1 Mean SCA (%) of entire Ladakh and the magnitude of trends

	Jan	Feb	Mar	Apr	May	Jun	Jul	Aug	Sep	Oct	Nov	Dec	Aut.	Win.	Spr.	Sum.	Ann.
Z-value	–0.63	0.63	0.42	0.14	–0.07	0.98	2.03	0.56	0.28	0.28	–0.14	0.00	0.00	–0.21	–0.28	0.98	0.56
p-value	0.53	0.53	0.67	0.89	0.94	0.33	**0.04**	0.58	0.78	0.78	0.89	1.00	1.00	0.83	0.78	0.33	0.58
Sen's S.	–0.33	0.24	0.18	0.02	–0.02	0.35	0.30	0.04	0.10	0.11	–0.06	0.08	–0.01	–0.10	–0.07	0.20	0.08
Lin. S.	0.04	0.19	0.05	–0.03	–0.04	0.23	0.33	0.09	0.04	0.25	0.12	–0.16	0.12	0.03	–0.01	0.22	0.09
Mean	64.26	70.62	68.22	59.26	45.10	29.64	16.73	13.32	20.57	28.58	37.25	48.63	28.93	61.17	57.53	19.90	41.85
SD	10.05	8.53	7.43	7.51	6.39	6.37	4.57	1.79	5.53	9.27	10.13	12.50	6.50	6.49	6.33	3.95	4.32
Trend	▼	▲	▲	▲	▼	▲	▲	▲	▲	▲	▼	▲	▼	▼	▼	▲	▲

Numbers in blue: Significance level, p-value < 0.05

Table D.2 Mean RSLE (m) of entire Ladakh and the magnitude of trends

	Jan	Feb	Mar	Apr	May	Jun	Jul	Aug	Sep	Oct	Nov	Dec	Aut.	Win.	Spr.	Sum.	Ann.
Z-value	0.28	−0.70	−0.91	−0.56	−1.26	−2.10	−2.35	−1.33	−0.21	−0.77	−0.28	0.28	−0.56	0.42	−1.19	−2.10	−1.12
p-value	0.78	0.48	0.36	0.58	0.21	**0.04**	**0.02**	0.18	0.83	0.44	0.78	0.78	0.58	0.67	0.23	**0.04**	0.26
Sen's S.	1.56	−2.59	−3.35	−1.10	−2.16	−5.20	−5.56	−1.29	−0.73	−2.67	−2.39	2.08	−2.08	0.56	−2.97	−3.53	−1.75
Lin. S.	−0.85	−1.42	−2.38	−1.24	−2.55	−5.13	−5.29	−1.40	−1.32	−2.80	−2.32	2.95	−1.92	0.23	−2.06	−3.94	−1.98
Mean	4832	4813	4958	5097	5247	5390	5526	5560	5478	5378	5212	5011	5353	4885	5100	5492	5208
SD	84.74	75.26	63.72	53.09	54.14	59.12	56.28	30.33	58.27	77.39	122.16	94.43	62.28	50.40	49.25	43.89	32.50
Trend	▲	▼	▼	▼	▼	▼	▼	▼	▼	▼	▼	▲	▼	▲	▼	▼	▼

Numbers in blue: Significance level, *p*-value < 0.05

Table D.3 Mean SCA (%) of the subregions and the magnitude of trends

	Jan	Feb	Mar	Apr	May	Jun	Jul	Aug	Sep	Oct	Nov	Dec	Aut.	Win.	Spr.	**Sum.**	Ann.
Siachan																	
Z-value	0.07	−0.63	−0.91	−0.56	−0.14	−0.07	1.26	0.49	0.21	0.14	0.28	−0.07	0.35	0.07	−1.05	0.63	−0.21
p-value	0.94	0.53	0.36	0.58	0.89	0.94	0.21	0.62	0.83	0.89	0.78	0.94	0.73	0.94	0.29	0.53	0.83
Sen's S.	0.05	−0.10	−0.26	−0.37	−0.10	−0.03	0.31	0.07	0.05	0.09	0.08	−0.04	0.04	0.01	−0.24	0.13	−0.04
Lin. S.	0.10	−0.09	−0.24	−0.38	−0.29	−0.19	0.27	0.05	0.00	0.10	0.10	−0.05	0.07	−0.01	−0.31	0.04	−0.05
Mean	81.55	81.76	78.63	74.62	67.48	57.45	46.35	41.85	51.53	58.63	64.34	73.57	58.25	78.96	73.58	48.55	64.81
SD	4.91	4.43	5.23	7.10	6.89	5.80	5.17	2.33	4.91	7.26	8.00	8.29	4.59	3.98	5.53	3.81	3.20
Trend	▲	▼	▼	▼	▼	▼	▲	▲	▲	▲	▲	▼	▲	▲	▼	▲	▼
Shayok																	
Z-value	−0.91	0.56	−0.63	−0.56	−0.56	0.90	1.40	0.21	0.49	0.70	0.14	−0.42	0.07	−0.35	−1.05	0.35	0.14
p-value	0.36	0.58	0.53	0.58	0.58	0.37	0.16	0.83	0.62	0.48	0.89	0.67	0.94	0.73	0.29	0.73	0.89
Sen's S.	−0.33	0.17	−0.21	−0.25	−0.19	0.07	0.27	0.01	0.11	0.45	0.18	−0.32	0.06	−0.10	−0.16	0.09	0.04
Lin. S.	0.03	0.04	−0.20	−0.24	−0.22	0.04	0.42	0.09	0.02	0.34	0.08	−0.29	0.13	−0.08	−0.22	0.18	0.01
Mean	65.32	67.70	64.63	58.53	48.38	34.56	18.16	13.46	22.10	32.42	41.42	52.44	32.12	61.82	57.18	22.06	43.26
SD	8.39	7.04	5.85	5.67	5.88	7.48	6.34	2.22	6.36	9.39	10.89	12.59	6.54	5.94	4.98	4.82	4.07
Trend	▼	▲	▼	▼	▼	▲*	▲	▲	▲	▲	▲	▼	▲	▼	▼	▲	▲
Pangong																	
Z-value	−0.35	0.70	0.28	−0.84	−0.28	−0.50	1.82	0.53	0.53	−0.21	−0.42	−1.05	−0.42	−0.91	−0.26	0.28	0.00
p-value	0.73	0.48	0.78	0.40	0.78	0.62	0.07	0.60	0.60	0.83	0.67	0.29	0.67	0.36	0.80	0.78	1.00
Sen's S.	−0.16	0.38	0.19	−0.26	−0.12	−0.04	0.21	0.04	0.16	−0.07	−0.26	−0.63	−0.18	−0.23	−0.01	0.04	0.02
Lin. S.	−0.14	−0.05	0.09	−0.21	−0.20	0.08	0.28	0.10	0.07	0.05	−0.08	−0.71	−0.03	−0.30	−0.11	0.15	−0.06
Mean	49.21	57.71	53.62	42.64	30.03	18.73	8.07	6.88	10.91	16.46	21.89	32.01	16.51	46.31	42.10	11.23	29.01
SD	14.81	11.97	12.62	8.03	6.72	6.44	3.90	2.22	6.02	10.44	10.65	12.92	7.24	8.94	8.15	3.44	4.82
Trend	▼	▲	▲	▼	▼	▼*	▲	▲	▲	▼	▼	▼	▼	▼	▼*	▲	▲

Indus																	
Z-value	–1.05	1.19	0.21	–0.83	0.00	1.86	1.61	0.63	0.00	–0.07	–0.42	–0.21	–0.14	–0.28	–1.07	0.63	–0.07
p-value	0.29	0.23	0.83	0.41	1.00	**0.06**	0.11	0.53	1.00	0.94	0.67	0.83	0.89	0.78	0.28	0.53	0.94
Sen's S.	–0.53	0.61	0.06	–0.16	0.04	0.10	0.17	0.05	–0.04	–0.07	–0.14	–0.15	–0.07	–0.12	–0.07	0.08	–0.02
Lin. S.	–0.19	0.31	0.10	–0.19	–0.14	0.08	0.23	0.06	–0.03	0.08	–0.04	–0.28	–0.01	–0.05	–0.07	0.12	0.00
Mean	44.78	51.83	46.42	35.03	23.08	13.61	5.06	3.76	7.79	12.62	17.90	26.71	12.86	41.11	34.84	7.47	24.05
SD	12.51	11.87	10.72	7.66	5.49	4.62	3.26	1.43	5.38	7.73	8.83	10.83	6.16	7.51	7.04	2.61	4.31
Trend	▼	▲	▲	▼*	▲*	▲*	▲	▲	▼	▼	▼	▼	▼	▼	▼*	▲	▼
Suru																	
Z-value	0.42	–0.14	–0.49	0.77	0.49	1.61	2.24	2.03	0.56	0.28	0.84	0.28	0.49	0.35	0.49	1.96	1.12
p-value	0.67	0.89	0.62	0.44	0.62	0.11	**0.03**	**0.04**	0.58	0.78	0.40	0.78	0.62	0.73	0.62	**0.05**	0.26
Sen's S.	0.16	–0.03	–0.14	0.32	0.15	0.83	0.59	0.21	0.19	0.13	0.78	0.18	0.30	0.21	0.15	0.52	0.24
Lin. S.	0.29	–0.06	–0.21	0.35	0.37	0.84	0.57	0.23	0.21	0.11	0.34	0.22	0.20	0.15	0.17	0.55	0.27
Mean	79.44	79.42	81.79	72.71	61.49	38.22	17.74	11.23	21.29	35.12	51.22	66.29	36.09	75.05	72.00	22.40	51.33
SD	11.48	7.74	4.95	8.24	8.03	10.07	6.90	2.91	5.82	11.10	17.20	10.40	9.14	6.59	4.72	6.37	4.11
Trend	▲	▼	▼	▲	▲	▲	▲	▲	▲	▲	▲	▲	▲	▲	▲	▲	▲
Zangskar																	
Z-value	–0.35	0.70	0.63	0.49	0.42	1.54	2.03	1.33	0.63	0.00	0.21	–0.07	0.00	0.21	0.42	1.68	0.77
p-value	0.73	0.48	0.53	0.62	0.67	0.12	**0.04**	0.18	0.53	1.00	0.83	0.94	1.00	0.83	0.67	**0.09**	0.44
Sen's S.	–0.31	0.20	0.17	0.21	0.22	0.63	0.37	0.09	0.12	0.12	0.10	–0.03	0.02	0.15	0.13	0.28	0.26
Lin. S.	0.16	0.38	0.24	0.19	0.19	0.56	0.35	0.18	0.11	0.59	0.24	0.01	0.28	0.18	0.21	0.36	0.27
Mean	72.93	80.89	82.44	72.05	49.89	26.37	11.61	8.34	15.65	24.62	37.46	52.50	26.10	68.77	68.13	15.44	44.56
SD	12.25	8.87	8.61	10.57	8.65	8.81	4.58	2.58	7.20	14.77	14.97	18.66	9.72	8.53	8.53	4.93	5.90
Trend	▼	▲	▲	▲	▲	▲	▲	▲	▲	▲	▲	▼	▲	▲	▲	▲	▲

(continued)

Table D.4 (continued)

Tshomoriri																	
Z-value	–0.35	0.14	1.05	0.00	0.35	0.84	1.89	0.56	0.63	–0.42	–0.49	–0.63	–0.14	–0.07	0.56	1.19	0.35
p-value	0.73	0.89	0.29	1.00	0.73	0.40	0.06	0.58	0.53	0.67	0.62	0.53	0.89	0.94	0.58	0.23	0.73
Sen's S.	–0.35	0.26	0.68	–0.07	0.21	0.31	0.25	0.07	0.14	–0.27	–0.23	–0.41	–0.03	–0.09	0.32	0.18	0.13
Lin. S.	–0.20	0.37	0.59	0.19	0.28	0.37	0.28	0.06	–0.09	0.22	0.17	–0.23	0.12	–0.02	0.35	0.24	0.17
Mean	61.27	75.13	73.45	61.13	34.05	14.70	6.34	6.21	12.02	17.67	21.60	33.72	17.18	56.71	56.21	9.08	34.77
SD	21.21	17.34	18.27	16.11	13.45	7.73	4.04	2.51	11.39	18.02	16.77	19.15	14.25	13.09	14.55	3.92	8.42
Trend	▼	▲	▲	▼*	▲	▲	▲	▲	▲	▼	▼	▼	▼	▼	▲	▲	▲

Numbers in green: Significance level, *p*-value < 0.1
Numbers in blue: Significance level, *p*-value < 0.05
* Modified Mann–Kendall Test

Appendix E
Snow Survey Forms

S. Passang, *Snow Cover Distribution and Dynamics*, Advances in Asian Human-Environmental Research, https://doi.org/10.1007/978-3-031-57692-8

Snow Survey information

Snow course No.: 7

Name: South Pullu

Sampler: Stanzin Passang Date: 07 January 2017

Station No.	Snow Depth cm	No. of Observations	Snow Water Equivalent mm	No. of Observations	Remarks
1	7	1	12	1	
2	7.5	1	11	1	
3	11	1	19	1	

Elevation in meter: 4771, 4813, 4860 Latitude: ________________ Longitude: ________________

SNOW SAMPLING: Started 11.15 a.m. Ended 1 p.m.

Sampling Conditions

Weather at the time of sampling: Temp. ____________ °C

______________ Clear

______________ Partly Cloudy

______x______ Overcast

______________ Raining

______x______ Snowing

______x______ Blowing

______________ Freezing

______________ Thawing

Snow conditions at Snow Course

___x___ Snow soft and powdery–not sticky.

________ Snow soft and wet-sticky.

Fig. E.1 Snow survey form a

____x____ Snow samples obtained easily.

________ Snow samples obtained with moderate difficulty.

________ Snow samples obtained with extreme difficulty.

________ Ice layer on ground. How thick? __________ cm

____x____ Ground frozen under snow.

________ Ground not frozen under snow.

________ Ground dry under snow.

________ Ground damp under snow.

________ Ground wet (saturated) under snow.

General Snow conditions

What elevation is snow-line generally? __________ m

Is snow melting on north and east slopes? __________

Is snow melting on south and west slopes? _____Yes_____

How many centimetres of fresh snow at snow course? __________ cm

Is there evidence of snow slides? _____No_____

Weather conditions of past month

________ generally overcast and stormy.

________ generally clear and cold

________ generally clear and melting.

Remarks: ____

____

Fig. E.2 Snow survey form b

Snow Plots	Snow Depth Plot-1 (4771m)	Snow Depth Plot-2 (4813m)	Snow Depth Plot-3 (4860m)	Remarks
Time Duration	11:15 AM to 11:30 AM	12:02 PM to 12:17PM	12:42 PM to 13.00 PM	
No. of Observations				
1	8	7.5	12	
2	12	8	12.5	
3	9.5	8.5	7.5	
4	10.5	9	11	
5	12	9	7	
6	14.5	8	10	
7	8.5	10	12.5	
8	12	10	10	
9	10	10	11	
10	12	11	10	
11	9.5	11	10	
12	14.5	9	9.5	
13	10	10	12	
14	10	9	11	
15	9	8.5	10	
16	10	9.5	10.5	
17	8	10.5	10	
18	10	10.5	15.5	
19	12	10.5	12	
20	10	11	10	
21	13.5	11	10	
22	9.5	9.5	10.5	
23	10	10	10	
24	10	10	15	
25	8	11	9.5	
26	9	12	13	
27	8	9	10	
28	10	12	10.5	
29	11	13.5	9.5	
30	11	9.5	10	
31	9.5	10.5	9.5	
32	10	9	9	

Fig. E.3 Snow survey form c

Appendix F
Khakshal AWS Build Procedure

S. Passang, *Snow Cover Distribution and Dynamics*, Advances in Asian Human-Environmental Research, https://doi.org/10.1007/978-3-031-57692-8

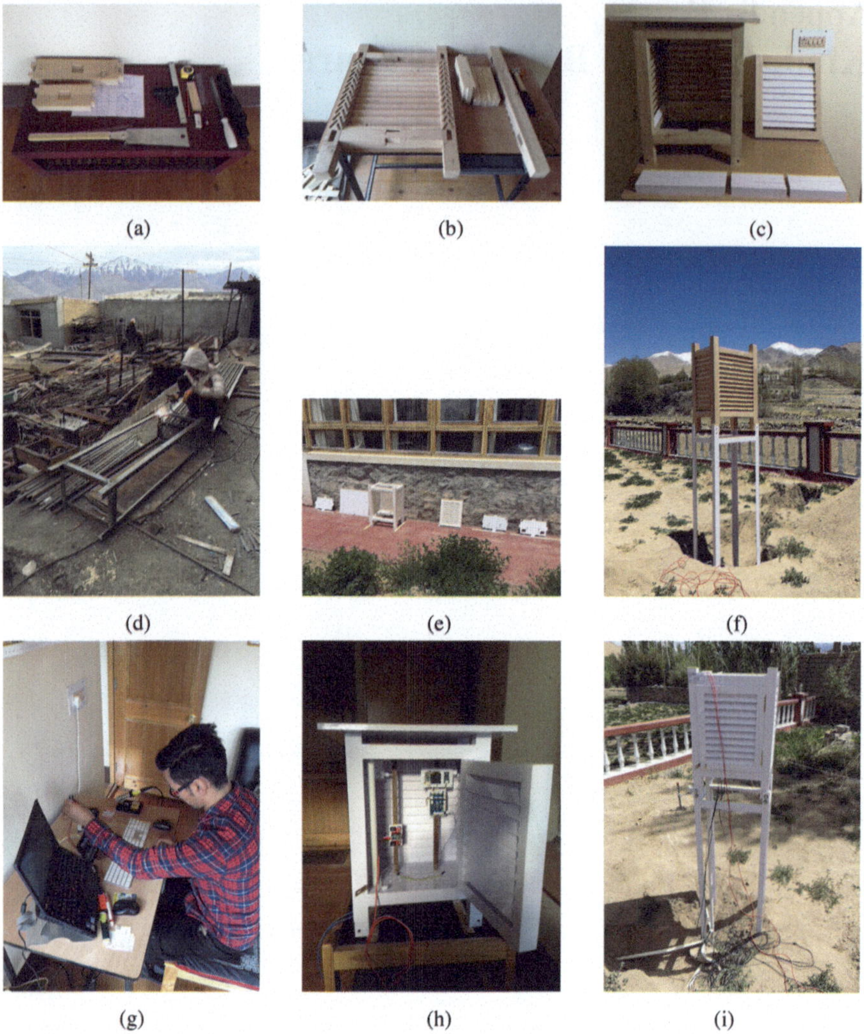

Fig. F.1 Khakshal automatic weather station build procedure I. (Photos: ©H. Thoma). (**a**) Tools and drawing. (**b**) Assembly I. (**c**) Assembly II. (**d**) Stand welding. (**e**) Painting. (**f**) Outdoor fitting. (**g**) Soldering/coding. (**h**) Indoor testing. (**i**) Electric wiring

Fig. F.2 Khakshal automatic weather station build procedure II. (Photos: ©H. Thoma). (**a**) Wind metres. (**b**) Camera. (**c**) Light intensity sensor. (**d**) UV sensor. (**e**) Calibration. (**f**) Camera box. (**g**) Rain gauge. (**h**) Summer. (**i**) Winter

Index

S. Passang, *Snow Cover Distribution and Dynamics*, Advances in Asian Human-Environmental Research, https://doi.org/10.1007/978-3-031-57692-8

SPRINGER NATURE

GPSR Compliance

The European Union's (EU) General Product Safety Regulation (GPSR) is a set of rules that requires consumer products to be safe and our obligations to ensure this.

If you have any concerns about our products, you can contact us on ProductSafety@springernature.com

In case Publisher is established outside the EU, the EU authorized representative is:

Springer Nature Customer Service Center GmbH
Europaplatz 3
69115 Heidelberg, Germany